Strategic Planning for Design Firms

by Raymond F. Kogan, AIA, and
Cara Bobchek

PUBLISHING

New York

This publication is designed to provide accurate and authoritative information in regard to the subject matter covered. It is sold with the understanding that the publisher is not engaged in rendering legal, accounting, or other professional service. If legal advice or other expert assistance is required, the services of a competent professional should be sought.

Editorial Director: Jennifer Farthing
Acquisitions Editor: Shannon Berning
Production Editor: Julio Espin
Production Artist: International Typesetting and Composition
Cover Designer: Kathleen Lynch

Published by Kaplan Publishing,
a division of Kaplan, Inc.

Printed in the United States of America

April 2007

07 08 09   10 9 8 7 6 5 4 3 2 1

ISBN 10: 1-4195-3954-X
ISBN 13: 978-1-4195-3954-1

Kaplan Publishing books are available at special quantity discounts to use for sales promotions, employee premiums, or educational purposes. Please email our Special Sales Department to order or for more information at *kaplanpublishing@kaplan.com*, or write to Kaplan Publishing, 888 7th Avenue, 22nd Floor, New York, NY, 10106.

Raymond F. Kogan, AIA, is the president of Kogan & Company, specializing in strategic planning and management consulting for design and construction firms. He has more than 30 years of experience in the practice, management, and marketing of architecture, engineering, and construction services. Ray is also a cofounder of A|E Advisors, a national network of specialized consultants serving the design and construction industries.

Cara Bobchek has over 20 years of experience working with and advising design and construction firms on communications, marketing, and strategy. She facilitates planning sessions and conducts original research on the architecture and engineering industries and has addressed numerous professional associations.

You can reach Ray or Cara through the Kogan & Company website at www.kogancompany.com.

Contents

A *c k n o w l e d g m e n t s*

The authors gratefully acknowledge:

Our friends and colleagues in the A|E Advisors network, whose knowledge is reflected throughout these pages

Our many clients who push, pull, challenge, and teach us with every engagement

Michael Tardif for his contributions to the manuscript

Our wonderful and supportive spouses, Suzanne Harness and John Reiter

BACKGROUND AND PREPARATION

1

WHY YOUR FIRM NEEDS A STRATEGIC PLAN

"A ship in port is safe, but that's not what ships are built for."
—Rear Admiral Dr. Grace Murray Hopper

Design firms—and by that broad term we mean architecture, engineering, environmental consulting, landscape architecture, planning, and interior design firms—are adept at thinking strategically about how to tackle a project; that's how they earn their fees. Think about taking on a design assignment for a client: you undoubtedly have a certain way in which you approach the work so that you're sure that you've listened to the client's needs, addressed those needs in a way that meets his or her schedule and budget for the project, and designed a solution that will help the client achieve his or her vision. In this book, we will show you how to leverage those same analytical, organizational, and problem-solving skills that you already possess as a design professional to envision, promote, and implement an exciting future for your own firm.

MOTIVES FOR STRATEGIC PLANNING

Perhaps business has been exceptionally good for you lately. Maybe, like so many design firms today, your firm has as much work as or even more work than it can handle. How could you justify taking the otherwise billable and immediately profitable time of your key people away from your valued clients in order to develop a strategic plan? The justification is simple: you will find it nearly impossible to maintain the status quo for your firm—much less to grow—over the long term without a plan.

There are some design firms that already know they need help and, therefore, are highly motivated to improve right away and plan for the future success of their business. In addition, many successful firms have already institutionalized the strategic planning process and take advantage of it regularly, enjoying the accomplishments that they have been able to achieve as a result. If you aren't one of the firms that routinely employs strategic planning as a management tool, we contend that you ought to be.

In our years helping design firms to develop strategic plans, we've found that firms decide to undertake a planning process for two basic reasons:

1. There is something—or some things—wrong that needs fixing.
2. There are opportunities that should be available to the firm that it is somehow missing.

In short, either something's wrong or something's not right *enough*. Most often, it's both.

Although firms' motives for strategic planning vary, there is a common thread among firms that do plan that we compare to the difference between being a passenger in a car and being the driver. When a firm's leadership doesn't have that secure feeling of being behind the wheel, it spells trouble, and the firm needs help. A firm's leadership may sense that there is more potential in the marketplace than the firm has the capacity or savvy to capitalize on. They may see evidence of operational or management problems within the firm. At the extreme, they may even feel the pressure of a threat so dire that it could lead to the end of their company.

Let's look at some of the more common scenarios that spark firms' interest in strategic planning:

- **Sensing opportunities to grow, but not being able to attain them.** If your firm isn't growing, then it's being left behind. In today's economy, opportunities are everywhere, but some firms find themselves unable to take full advantage of the boom times. Their constraints may have to do with marketing, project management, staffing, geographic limitations, or a combination of these factors. Clients, employees, and even prospective employees want to work with a winner, and a stagnant firm is unappealing across the board.

- **Difficulty in recruiting and high turnover.** Design firms find themselves competing actively for employees as aggressively, if not more so, as they compete for work. There may be aspects of the culture of your firm that cause it to attract or lose valuable talent.

- **Competition.** When you face losing a share of your market to competitors, you must improve your marketing program, demonstrate to your clients that you truly understand their needs, and improve the way you execute and deliver your projects.

- **Waning profits.** The median profitability of all design firms has risen rapidly in recent years to a 2006 median profit margin of approximately 14 percent of net revenue. If your firm isn't achieving its profit potential, what do you need to improve in order to do so?

- **Ownership transition.** Almost all design firms started small with an ownership structure (often a single owner or small group of equal partners) that may no longer be appropriate or even viable. The financial impact of an ownership transition can sometimes present a make-or-break situation for the future of a firm.

- **Leadership succession.** Above all other issues that a firm faces, transitioning from one leader—in many cases the firm's founder—to the next generation of leadership must be well thought out and executed far in advance of the transition itself in order to succeed and therefore to perpetuate the firm.

CONTROLLING YOUR FIRM'S DESTINY

At any moment in time, your business is either on an upward or downward trend. But you have a choice when it comes to both the near- and the long-term future of your firm: you can try to shape it or you can just let it happen to you. The strategic planning process is an opportunity for you to *take control of the destiny of your firm*. You can *decide* what you want your firm to be, visualize its potential, understand the circumstances in place today that could stand in the way of your achieving that potential, and devise and then take the steps necessary to achieve the future you desire.

THE PROCESS, COMPONENTS, AND SUBSTANCE OF STRATEGIC PLANNING

This book serves two purposes: it is an introduction to strategic planning for design firms that have never engaged in the process, and it is a guide for improving the process and the results for those firms that have conducted strategic planning.

- **Part One: Background and Preparation**
  In Chapters 1–6, we discuss the *background* and *preparation* for strategic planning, which involve convening a strategic planning committee, gathering information, and dedicating a retreat to developing a plan.

- **Part Two: Components of Strategic Planning**
  In Chapters 7–12, we describe the six sequential *components* that make up a strategic plan.

- **Part Three: Strategic Planning Issues**
  In Chapters 13–20, we discuss the substance of some of the more common *key issues* that many design firms address in their plans.

Although we describe the six components of a strategic plan in detail in later chapters, let's take an introductory look at them here:

- The **mission** (Chapter 7) expresses the essence or purpose of a design firm. The mission statement gives people inside *and* outside the firm a common understanding of the firm's reason for being.
- The **vision** (Chapter 8) is a big-picture, long-term description of what a firm aspires to become at some specified point in the future.
- **Key issues** (Chapter 9) refer to the major obstacles that will hinder a firm's ability to achieve its vision.

- A firm's **goals** (Chapter 10) are the midterm (shorter than the long-term vision) quantitative, measurable targets that the firm sets and toward which it will gauge its progress and for which it will be accountable.
- **Strategies** (Chapter 11) are the ideas the firm develops to address its key issues and reach its goals.
- **Action plans** (Chapter 12) are the short-term (often one year) specific "tactical" tasks that the firm will implement in order to put its strategies into motion.

What Strategic Planning Is Not

Many design firms believe in the concept of strategic planning, and they believe that they even engage in it. But they also may hold a number of misconceptions about strategic planning. A strategic plan is *not*:

- **A marketing plan.** Most firms produce marketing plans annually, describing their target markets, setting a marketing budget, and forecasting anticipated revenue. Strategic planning includes marketing because obtaining work is the lifeblood of any design firm. But strategic planning is much broader and more holistic than just marketing and can include leadership and management, organizational structure, staffing and human resources, financial management and performance, project delivery, ownership transition, and many other topics.
- **An annual plan or budget.** Many design firms develop an annual budget, forecasting their revenue for the coming year and budgeting line items for anticipated expenses. But strategic planning encompasses much more than just numbers. It starts with big-picture thinking, sets a direction and destination for the firm, and lays out a road map for how to get there.
- **An annual management retreat.** It's worthwhile for firm leaders to get away from the office to an off-site event to discuss matters of importance, but the notes from these meetings often describe a nonprioritized series of discussions, many of which deal with items at a level of detail that, though important, would not be

considered "strategic." A strategic planning process also involves an off-site retreat, and, properly facilitated, it consists of an intense, sequenced series of group activities that results in a *product*, a concise planning document to be used as a real management tool to guide the firm.

- **Solely a team-building exercise.** Getting away from the office to bond with coworkers can be valuable and even enjoyable, but developing a strategic plan as a group is akin to a barn-raising: a group of people work hard together as a team and have something special and useful to show for it when they finish the work.

THE BENEFITS OF A STRATEGIC PLANNING CULTURE

The best firms to work *for* (from the employee's perspective) are usually also the best firms to work *with* (from the client's or consumer's perspective). Witness perennially top-rated firms listed among *Fortune* magazine's annual "Best Companies to Work For" citations such as Southwest Airlines, Google, or The Container Store. When you read about the top firms to work for in any industry, you'll find that the winners offer far more than just generous benefits such as on-site day care and every third Friday off. The deeper culture of a firm is critical to developing and maintaining the kind of dedication and motivation among employees that often characterize the firm's success. The exceptional firm's people look forward to coming to work, feel rewarded for their innovation and performance, and see their career path at the firm. They are proud to be part of their company, and that is part of what makes their company so competitive. Firms that use strategic planning are more likely to be able to foster that kind of environment. Just having a strategic plan at all is indicative of a progressive culture of success. The strategic plan:

- Paints a vivid picture of a firm's future
- Lets everyone understand the firm's direction and priorities

- Is developed by top management with ideas and input from a broad base of the firm's employees, so it inspires confidence, displays leadership, and instills a sense of direction that staff in design firms crave
- Is tangible, so it feels "real," and sharing it throughout the firm engenders trust and support
- Provides a common backdrop against which a firm can make decisions and to which employees can hold each other accountable
- Can be used not only to help a troubled firm get back on track but also to help a good firm become stronger than ever

Simply put, firms that know where they want to go are much more likely to get there than firms that don't plan their journey.

Are Things Changing Too Much, Too Quickly to Plan?

We read an article recently by a national business leader who posited that strategic planning constrains companies and hampers their growth. Referencing the "bewildering pace of change and the blizzard of activity [that] can make planning difficult in the New Economy," this business leader claimed that rapid changes in the world prevent planning. But we know that strategic planning can be an effective way to address these kinds of changes.

In another article, a construction industry firm leader wrote his opinion that "business is changing far too rapidly to predict where we will be, what we will be doing, and what we will look like in a year, much less in three." The design and construction industry may indeed be changing rapidly, but other industries are even more dynamic. Look at information technology as one example. The business leaders of successful firms such as Microsoft and Google are not at all bewildered by the mergers, acquisitions, regulatory changes, and market conditions that are constantly "de-leveling" their playing field. Because of the huge stakes involved, these companies routinely develop purposeful strategies that are integral to their carefully thought-out business plans.

Our design and construction marketplace is absolutely dynamic. Strategic plans need to be flexible. But setting a destination, plotting a course, and implementing action plans make reaching your goals much more certain than casting your firm's fate to the wind.

Think of planning for your firm's future as planning for a family vacation. You can start thinking about it as you're pulling out of the driveway. Or you can decide where you're going to go well in advance. If you plan, you'll know the best route to take to get there, what to pack in order to be prepared, what you'd like to see along the way, and where to stop.

The Strategic Plan As a Recruiting Tool

Issues related to finding people—and the right people—have been and will probably continue to be the biggest challenges holding design firms back from their greatest potential. We're all familiar with the age-old question that we usually ask of job candidates: "Where do you want to be in five years?" In today's "seller's market" for design professionals, candidates are in the position to ask that same question of the firm with which they are interviewing. Imagine an in-demand job candidate interviewing with your firm and also with a competitor. At the interview with your firm, you can proudly show that you know exactly where you want your firm to be in five years and what it will take to get there because you have a strategic plan. If, at your competitor's office, the savvy candidate asks a similar question and the response is that the "founder has it all worked out in his head," which firm do you suppose the candidate would rather work for? Which firm would you rather work for?

What Employees Want in a Strategic Plan

A recent "*USA Today* 'Snapshot'" from the front page of the business section illustrated the findings of a nationwide poll asking employees the open-ended question "What aspect is important when your employer communicates with you?" Among the top responses were "Showing how to fit into the company's vision," cited by 47% of respondents and "Explaining the company's vision," cited by 45% of respondents. Strategic planning satisfies both of these needs, especially when everyone at the firm is prepared for it.

2

LAYING THE GROUNDWORK

"Make no small plans; they have no magic to stir men's blood. Make big plans. Aim high in hope and work."

—Daniel Burnham

Successful strategic planning depends a great deal on the attitude and commitment of the participants in the process, beginning with the leader in charge of your firm. If the participants can imagine the firm's potential and are also willing to see the firm as it really is and squarely confront challenges and obstacles, the firm is more likely to develop a strategic plan that will really take it places. On the other hand, there are some characteristics of a firm's culture that can undermine even the most well-intentioned strategic planning process. It is important to cultivate an environment in which the strategic planning process can thrive and drive a firm toward the successful future it desires.

If your firm is prepared for strategic planning, it will also be better poised to successfully manage the changes that may be called for in the plan. In Chapter 5, we discuss the importance of sharing the newly minted plan with all of a firm's employees to build broad-based enthusiasm and buy-in for its direction and strategies. Just as one key to a successful plan is sharing it with employees after you develop it, it is also important to prepare them for the planning process beforehand.

HALLMARKS OF A STRATEGIC PLANNING CULTURE

Once a firm has institutionalized the process of strategic planning, it will find itself with an ongoing "bias for action," an attribute that Tom Peters and Robert Waterman identified as one of the keys to success in their iconic business book *In Search of Excellence.* But if a firm has not undertaken strategic planning in the past or has tried it unsuccessfully, it is helpful to first lay some groundwork.

In our experience working with design firms, we have found that the firms most successful at strategic planning often possess the following characteristics:

- **Their leaders are inspired and inspiring**. The CEOs of successful design firms push themselves and their people to change and improve continuously on many fronts, and that philosophy permeates the entire firm. They understand that strategic planning has the ability to move the firm ahead, even if the firm is already doing well.
- **They consider and lead their own firms with the same level of sophistication as their clients' companies**. Despite the rapidly growing popularity of strategic planning for design firms, the practice is still more commonplace in the general business world—and even among institutions and public-sector organizations—than it is among design firms. It is more than likely that your best clients have strategic plans that they follow. They consult the experts in their industry, stay abreast of changes, anticipate trends, and make strategic investments in order to be successful over the long term. In a competitive environment, why should your own design firm settle for any less?
- **They are willing to look at their firms from the outside as well as from the inside**. Some individuals have an ability termed "situational awareness," defined as knowing and understanding what is going on around them and sensing how things will change—in other words, being coupled to the dynamics of their environment. Firms that are open to the tenets of strategic planning tend to be

those that make it their business to be connected to the marketplace, their client community, and their profession. They not only keep their clients, friends, and partners in the profession informed about their progress, but they also listen to their clients' business needs and stay on top of the trends in their clients' industries and the business world at large. They read voraciously. Periodically, they query clients formally or informally about industry trends and the design firm's performance. They participate in their own architectural, engineering, planning, environmental, or other professional associations, as well as in their *clients'* business and professional organizations. They regard maintaining their firm's "situational awareness" as a strategic investment.

- **They foster respect throughout their organizations and emphasize candid communication.** These firms do more than hold an occasional "all-hands" meeting, have their marketing or human resources group crank out an in-house newsletter, or keep their intranet up to date. They truly trust their management, who in turn trust their staff, and they treat their employees like responsible adults. They value honest communication at all levels, and they make it easy to participate in candid dialogues. These firms might convene a young professionals' committee, for example, that meets regularly to generate ideas and advise management on issues germane to recruiting, retention, mentoring, and training, and they might also conduct annual employee climate surveys. They make sure that management is not only accessible to staff, but also invites meaningful interaction with the staff, and they conduct frank and constructive performance reviews *often* and *on time.* They recognize that open communication is a two-way street, a flow of information that benefits everyone.

- **Because they respect the maturity and intellect of their staff, they share financial performance information throughout their firms, from top to bottom.** Design firms that are fertile ground for successful strategic planning are good at *reporting up*—about project and people performance to management—as well as *reporting out*—about the firm's performance and prospects to the entire staff. A staff that is informed about the firm's performance will understand, as well as come to expect, the need to plan strategies to improve that performance. When a firm explains its financial performance metrics to the staff, it helps everyone to understand what is causing the firm to reach—or miss—its profitability goals, and it allows them to connect their own daily behavior to the firm's performance. When employees are exposed to the business of their firm, they become more effective employees and managers, they build their leadership potential, and they are better prepared to embrace strategic planning as a tool for improvement.

- **They understand the connection between strategic planning and marketing**. Although it is true that a strategic plan is not equivalent to or a replacement for a marketing plan, there nevertheless tends to be a correlation between firms that are effective at marketing planning and firms that embrace strategic planning. Firms that are effective marketers make it their business to understand their clients' businesses and issues so that they can solve the clients' problems. Strategic planning requires turning that same focus inward on your own firm to understand your business, establish your vision, and address the issues that are standing in the way of your firm's achieving its goals.

This list of attributes is a tall order, and, frankly, few firms possess them all. But if your firm does not share enough of these characteristics, embarking on a strategic plan can help you address them and integrate them into your firm's culture.

Roadblocks to a Strategic Planning Culture

In order to develop and carry out a successful strategic plan, your firm has to be culturally prepared for planning. In our experience, the following attitudes—often displayed and even proclaimed by firm leaders—can present roadblocks to strategic planning success:

- *The market is good, and we're making money, so whatever we're doing has been working for us so far.* Sure, plenty of firms have done pretty well for themselves in times when clients are flush and the economy can sustain a construction boom. When your firm has been riding the wave, it's easy to confuse what has been "working for you" with what has been "working *out* for you." But it's risky to leave your firm's future to the will of external market forces. Strategic planning is taking control of your firm's destiny, not leaving its direction up to the whims of the marketplace.

- *Nothing seems to be broken, so we don't need to fix anything.* The thought of making major organizational or operational changes can be daunting to anyone. People want to know their place within a firm, and anything that could alter that can be scary. Chances are that your firm will not embark on a major organizational restructuring as a result of a strategic plan, but you must nevertheless be open to change and improvement if your plan is to be successful. The saying "no pain, no gain" applies to exercise and also to the efforts of a firm that change it for the better.

- *It's impossible to plan far ahead in a dynamic marketplace.* It's not impossible at all; in fact, it's necessary if your firm is to be competitive. Successful design firms—probably including your competitors—conduct research and use it to position themselves on an ongoing basis.

- *We're too busy with the "real" work of the company to take time to plan.* Strategic planning isn't executive busywork. If you neglect planning for your future because you are busy today, you are bound to miss out on opportunities to remain successful. Of course, the planning process itself does take some time, as does implementing the actions to achieve your goals, and this time isn't billable. You'll need to take a longer view of your firm's success to hold people accountable for the actions they have been assigned. For example, if one factor in your firm's incentive compensation program is keeping billable hours as high as possible, the conflict of seemingly urgent priorities versus the long-term importance of plotting the firm's direction could undermine your ability to implement the plan.

- *We already have a marketing plan and financial goals.* A strategic plan addresses important areas of your business beyond marketing and revenue goals. Focusing too narrowly will hold your firm back from having a compelling vision to move forward.

- *Strategic planning is too "corporate" and involves too much paperwork.* You'll be surprised: good strategic planning should be streamlined and efficient, and it needn't take more than a few weeks or months to develop the plan, from start to finish. And as you'll see in this book, we believe that the best strategic plans are straightforward and slim—perhaps no more than ten pages in length.
- *A strategic plan could be a threat to my position.* It doesn't have to be! Think of the strategic planning process as an opportunity to collect valuable information and ideas that will make you a stronger and even more successful and respected leader.

PREPARING FOR STRATEGIC PLANNING

Over and over, employee surveys and management interviews reveal to us that one thing staff members crave is more communication from upper management. In fact, design firm staffers seem to have an insatiable appetite for information about their firm. Preparing for strategic planning is a perfect opportunity for a firm's leadership to get more personal with the rest of the employees. Meet with the management and staff and travel to other offices, if your firm has them, to explain that your firm is about to undertake an important process that will define an exciting future for the company, and show your own enthusiasm for it. Explain the process and reinforce that everyone in the firm will participate on some level—and benefit at all levels—whether in interviews during the information-gathering phase, through a firm-wide employee survey, or with a responsibility to carry out an action that results from the plan. The objective is to connect the individuals to the process. This is a great opportunity to make personal contact, answer questions, and let people know about strategic planning:

- **Tell people what to expect.** Describe the planning process, including who is participating and how. Develop a schedule for the planning activities, publish it, and then stick to it.
- **Reassure your staff that you appreciate their candor.** Model that you are willing to listen to everyone's thoughts, regarding both

good and bad news. Emphasize that the firm must understand exactly where it stands today and what issues it faces for the future.

- **Encourage participants to think strategically.** Get them to look beyond their immediate purview—their office, department, studio, or team—and see the firm and its future as a whole. Just as a rising tide raises all ships, what benefits the entire firm benefits everyone in the firm.

- **Make sure the message comes from the top.** Be an enthusiastic and engaged champion for the strategic plan. This is also important to keep in mind if your firm is working with a consultant to develop the plan. If the firm leader steps back from the process after the consultant is hired, this can send the message that the CEO is simply "buying services" and isn't really behind the plan.

- **Understand that you need the commitment of the entire staff in order to be successful.** Ask for their support, and mean it!

The planning process should shine a light on all aspects of a design firm, including leadership and management, organizational effectiveness, marketing and business development, project management and delivery, financial management, human resources, and technology. Once you have laid the groundwork and opened your management's and staff's minds to a strategic planning culture, you can begin to gather the information on which to base your plan.

3

GATHERING INFORMATION

"As a rule, he or she who has the most information will have the greatest success in life."

—Benjamin Disraeli

A good strategic plan is based in reality. That means that the information on which the plan is based must be valid and accurate. Just as undocumented, out-of-date, or incorrect assumptions can be disastrous to a design process, they can also threaten a strategic planning process.

Before developing a strategic plan, you need to gather information about your firm that is relevant, accurate, and current. This information will help you identify the strengths of your firm that you will want to preserve and build around, as well as the issues facing your firm that may be holding it back, which the strategic plan will help you overcome.

There are several kinds of information that you can collect and combine into a coherent form that you can use for thoughtful decision making. Of course, you will need to gather quantifiable data, such as statistics and financial performance indicators. Just as illuminating, if not more so, is the variety of qualitative information that you will discover to serve as the basis for your firm's strategic plan.

Combining quantitative data with qualitative information reveals several things about your firm. One is an objective and accurate picture of how your firm is performing compared to past years and compared to the performance of similar firms in the industry; another is a sense of how others within and outside your company view your firm, which is often quite different from what you expect. All together, this information will enable you to recognize that change and improvement are necessary—the first and most important insight of the strategic planning process.

QUANTITATIVE DATA

In order to have a quantitative baseline for strategic planning decisions, firms can review financial indicators and related data. This type of analysis is diagnostic, much like going to the doctor for a blood test. The results come back in quantitative form for each indicator—"good" cholesterol, "bad" cholesterol, triglycerides, etc.—and you can note your own results and compare them to an acceptable range for healthy people. The doctor then uses these findings to determine your relative situation and prescribe ways to improve it.

When analyzing your firm, you should note the pattern of performance over a sufficiently long period of time—the past four or five years, for example—to discern trends, as well as compare key indicators to those of other firms similar to yours to gain a sense of where your firm stands among its competitors in the marketplace. Avail yourself of the benchmarking data that are commercially available for design firms in the form of annual surveys and reports published by organizations that keep statistics on the design industry. These surveys encompass a wide range of metrics that go far beyond basic financial performance measurements and also include compensation, human resources, marketing, and information technology. In addition, helpful background statistics may also be available on the Web from federal and local government sources, such as the Department of Labor Statistics and the Census Bureau.

It is important to examine your firm in two ways: looking at trends over time and comparing it to other firms via statistics on industry medians and upper and lower quartiles of firms of similar size, markets, and services. Sometimes firms assume that they are doing as well as is possible until they compare themselves to a statistical base of other design firms, only to discover that many other firms in the industry are doing much better.

You'll need assistance from your firm's accounting, marketing and business development, and human resources experts to complete the analysis. Firms should analyze any and all of the following kinds of data in order to identify key issues that they may need to address in the strategic plan:

Financial Performance

Growth

- Sales, gross revenue, and net revenue (the gross revenue received by the firm, less fees paid to subconsultants and project reimbursable expenses) trends over the last five years
- Total (full-time equivalent) employee numbers over the last five years
- Backlog trends, measured in months of net revenue

Profitability

- Operating profit margin (pretax, predistribution) on net revenue: the percentage of net profit/loss before taxes and discretionary distributions (bonuses, incentive compensation, profit sharing, etc.), based on net service revenue
- Return on equity: the rate of return of the owners' investment in the firm, after taxes and bonuses; calculated by dividing a firm's after-tax profit/loss on net service revenue by total owner's equity
- Return on assets: the rate of return on the firm's assets; calculated by dividing after-tax profits by total assets

Operating Efficiency

- Net direct labor multiplier: an indicator of project-related staff efficiency calculated by dividing net service revenue by *direct* labor dollars (the dollar amount of wages and salaries directly charged to projects, excluding fringe benefits)
- Net payroll multiplier: also known as the "revenue factor," it measures how efficiently an entire firm generates revenue; calculated by dividing net service revenue by *total* labor dollars
- Utilization rate: also known as chargeability, it measures direct labor dollars as a percentage of total labor dollars

- Net revenue per total staff: calculated by dividing the firm's net service revenue for a given period by the average number of full-time equivalent staff that the firm employs during the fiscal year
- Overhead rate (before incentive/bonus): the ratio of total overhead (including general and administrative overhead, indirect labor, and payroll expenses, but excluding bonus and pension provisions) to total direct labor
- Overhead rate (including incentive/bonus): the ratio of total overhead (including general and administrative overhead, indirect labor, payroll expenses, and bonus and pension provisions) to total direct labor

Operating Liquidity

- Current ratio (excluding deferred taxes): current assets divided by current liabilities
- Financial leverage (interest-bearing debt divided by stockholders' equity)
- Total debt to equity: total liabilities divided by total equity
- Average collection days (receivables outstanding): average number of days it takes to collect accounts receivable from clients, measured from the time an invoice is entered into accounts receivable to the time it is credited against accounts receivable; calculated by dividing gross annual revenue by 365 (to get average daily revenue), and then dividing accounts receivable by average daily revenue

Marketing Performance

- Marketing expenses as a percentage of revenue
- Marketing success in terms of "hit rates," both from proposal to short list and from short list to win

Recruiting and Retention

- Recruiting success statistics including resumes received, hires made, and average length of time a position remains "open"
- Overall turnover rate—both voluntary and involuntary—trends over an appropiate period

A note of caution: aiming to align your firm with a published industry median is not necessarily the best course of action. This may be obvious when considering performance-related indicators such as profitability or other financial measures, cases in which every firm would like to be "above average." But when it comes to making investment decisions about the firm, *investing more* than the industry median in marketing should yield *better than industry average* marketing results, and likewise for human resources, accounting, technology, and other areas of comparison.

QUALITATIVE INFORMATION

Although quantitative data is an important foundation for strategic planning, there is no substitute for a realistic and accurate understanding of your firm as seen through the eyes of those who are most important to its future success: your managers, employees, and clients. The best way to gain this understanding is to ask for it directly—through one-on-one interviews with key personnel, small focus groups, broader employee surveys, and individual client interviews.

Reach Out by Reaching In

Strategic plans are most successful when everyone in your firm feels in some way involved in the process and invested in the results. The information-gathering phase is the ideal time to reach out to as many of your employees as possible—if not to all employees—in order to get an accurate internal picture of issues that your firm must address through strategic planning. This will also build good will within your firm and give everyone the feeling that they have a stake in the plan's—and their firm's—success.

It is important to look for the patterns and trends that are revealed throughout the course of key personnel interviews and in responses to employee surveys. These will point to key issues that you may need to address in your firm's strategic plan. Yet it is also important to recognize that it is common for employees at all levels to dwell on the negatives when given the opportunity to "vent" during one-on-one interviews or in an anonymous response to a formal survey. Although a pattern of similar critical responses may indeed indicate a legitimate problem that strategic planners need to address, there will always be an "outlier" here and there among the responses—someone's personal gripe—that should not distract from the truly key issues.

The best way to ensure that interviews and surveys are useful qualitative tools is to thoughtfully prepare a series of even-handed and open-ended questions designed to elicit both the positive and the critical perceptions of employees. Provide anyone being interviewed with the questions in advance so that they can consider their responses and prepare notes. In the case of a broad survey, be sure to provide enough time to allow employees to set aside 20 to 30 minutes out of their schedules to respond thoughtfully and clearly; allowing a two-week period before a deadline should be adequate. Any longer could result in the survey request moving to the bottom of an in-basket or falling off the bottom of a screen of unanswered e-mails, while any less time could contribute to rushed, less thoughtful responses.

The Value of an Outsider

Many firms find it wise to engage an objective third-party strategic planning expert to assist in information gathering. Collecting, compiling, and analyzing all of the relevant information can be time-consuming; bringing in outside help not only assures that this important phase of the process remains on track and is not put on the back burner due to urgent firm- or project-related priorities, but it also lends a critical measure of objectivity, credibility, and a willingness to deliver the difficult messages that often isn't possible from a firm "insider." In any case, it is imperative that everyone who participates in a one-on-one interview or a broader employee survey is assured that their comments will be anonymous and confidential. The goal is for everyone to feel comfortable sharing their most candid observations—good or bad—about the firm without any concerns about censorship or retribution.

Ask Key Personnel

At a minimum, interview the top managers of your firm prior to the strategic planning sessions. After all, many—if not all—of them will be involved in developing the strategic plan itself, so their individual and collective insights are crucial. Presumably, the top managers possess higher-level strategic perspectives on the firm as a whole, as well as on individual projects, departments, clients, or people. But to really get the big picture about your firm, you also need to interview other key personnel whose insight you might not routinely seek at the strategic level. Their perspective and experience "in the trenches" is necessary to understand what is really happening at your firm. Ideally, a firm will interview a cross section of managers and technical and nontechnical personnel from representative departments and offices—a vertical or diagonal "slice" through the firm—in preparation for developing the strategic plan. The number of people interviewed should be enough to gauge patterns in their responses, but not so many as to encumber the process. Increasing the number of people being interviewed typically generates a diminishing rate of return in terms of cost- and time-effectiveness regarding the findings from the interviews. In other words, though the 30th, 40th, and 50th interviews take just as much time, effort, and expense as the first 10 or 20, it is less likely that they will yield new findings, as long as those 10 or 20 represent a true cross section of the firm.

Good questions that are thought provoking and inspire people to think strategically include the following:

- What would you like the firm to achieve in the future? (Describe future achievements in terms of any or all of the following: image and reputation, geography, markets served, services provided, size, management and organization, operational systems, culture, or any other attributes that come to mind.)
- What strengths does the firm currently possess that will allow it to achieve these results?
- What obstacles or problems must the firm overcome that could hamper its ability to achieve these results?
- What makes your firm special, or different from or better than other competing firms?
- What would you like the firm to accomplish through this strategic planning process?

One-on-one and focus-group interview sessions allow the interviewer and interviewees to engage in an intimate dialogue and give the interviewer the opportunity to probe more deeply any subjects of particular interest. When the interviews are complete, the third-party interviewer (often a strategic planning consultant) should compile the responses into a report that reflects the themes and patterns that point to key issues that the firm will address in its strategic plan. This report warrants careful reading and consideration by the strategic planning team in advance of the strategic planning retreat sessions in order to gain consensus on the interpretations of the comments and the strategies to address any key issues indicated.

For example, consider the following representative comments provided by employees of a large engineering firm:

"Nobody really understands our organization. Our organization charts aren't published at all. Even regional managers don't have a copy of our org chart."

"The organization can be confusing. Who's my boss? It would support good quality work if the discipline leaders overseeing the work were clear."

"Our newer people tend to come from places with more structure, while we've always been pretty loose about our organizational structure. There's the 'real' organization chart and the 'black market chart' of how things really work."

"I have three supervisors: Ed, Jerry, and Cindy."

Together, these comments seem to indicate a problem with reporting structure. The firm would have to determine if the key issue indicated by these comments is simply a matter of internal communication (i.e., this information exists but people don't know about it) or if there is truly a problem of not having established clear reporting structures.

Ask All Employees

No matter the size of the firm, a survey of all employees is usually feasible and often advisable. For cost-saving reasons, larger firms may choose to survey a random sampling of employees, but they should consider including all employees. The good will built through allowing everyone to have a voice often outweighs the effort of tabulating a large number of responses. This is especially true when the method of conducting the employee survey takes full advantage of the efficiencies of current technology, such as using Web-based survey tools.

In order to turn up key issues that might need to be addressed in a strategic plan, employee surveys should include a mix of multiple-choice or rated questions, where statistics representing the responses can be compiled easily and compared, along with a few open-ended questions. Good questions that have proven to elicit responses relevant to strategic planning cover the following:

- The employee's career at the firm, such as satisfaction with types of projects assigned, level of responsibility, training opportunities, salary, and benefits
- The employee's perception of the firm's performance in the marketplace and its reputation among clients and competitors
- The firm's internal operational effectiveness in project management, technology, administrative support, and marketing support
- The employee's opinion as to the best thing about the company that the employee would not want to see changed
- The employee's opinion about the most important thing for the company to improve

How frequently an issue is raised in proportion to the number of employees who respond to the survey gives a sense of the issue's relative importance to the employees; however, a firm must carefully consider the source of the comments. For example, lack of profitability might be absolutely the most critical issue that a firm is facing, but a firm-wide employee survey is often less likely to show that, because in many firms, employees typically aren't aware of detailed profitability metrics (even if they should be). Therefore, financial performance is a topic that might be more likely to come up among a group of senior managers.

Ask Clients

To get a sense of how your firm is perceived in the marketplace, there is no source better to ask than the marketplace your firm serves. Consider contacting the following:

- Current clients with whom you have an ongoing relationship
- Past clients for whom you have previously worked but who have since chosen to work with a different design firm
- Prospective clients, those for whom you would like to work but as yet have not

Any firm can take advantage of the opportunity to talk to its clients in preparation for strategic planning, and all of them should. In our experience conducting dozens of surveys of a variety of design firms' clients, we have found that current, past, and even prospective clients are willing and happy to take 15 minutes to talk about their perceptions of the design firm, their take on market trends, and even their own organization's goals for the future. Firms can engage clients in this way through telephone, written, or online surveys. Especially with an objective third party, clients are comfortable delivering kudos as well as candidly relating any bad experiences from which the firm can draw conclusions.

Be sure to reach a good cross section of clients by type, region, and market. Depending on your markets, you might contact owners, end users, or developers of projects, or other consulting firms involved in your industry that are your clients. Don't forget strategic teaming partners and subconsultants, if they are relevant to your business.

Remember to Say "Thanks!"

Regardless of whether someone participated in interviews of key personnel, a firm-wide employee survey, or a client survey, the job isn't complete until the firm has thanked the participant for his or her valuable input. In the case of key personnel interviews, a thank you may be in person or via e-mail, largely depending on the size of the group being surveyed. In the case of a firm-wide employee survey, everyone should receive a memo of thanks from the firm leader (because participants are typically anonymous, participants can't be thanked individually). In the case of a client survey, all participants should receive an individual business letter from the leader.

In each case, in addition to thanking the participants, the communication piece should summarize in a brief but meaningful way the key messages learned from the surveys or interviews and what the firm intends to do in order to address them. Letting the participants know that the firm is going to take action to address what it learned makes them feel that the time and thought that they put into the process was worthwhile.

In addition to providing invaluable information on which to base strategic planning decisions, a client survey provides plenty of other benefits. Simply undertaking the client survey lets the participants know that your firm cares about their opinions. It also provides the opportunity to identify problems on specific projects that the client might not feel comfortable discussing with the project manager but feels free to share with an independent third party. However, this high-level strategic client survey should not be thought of as a substitute for end-of-project or periodic "report card" surveys that you might (and should) conduct with your clients. As with employee surveys, the greater value is in the patterns and trends—not as much in the outlier comments—that emerge from asking multiple clients the same questions. And though it is critically important for the client contact not to perceive this type of survey as a sales call in any way, you might nevertheless learn of opportunities through the interviews. At the very least, you will build good will and positive public relations through a client survey; the following direct quote from a recent client survey that we conducted on behalf of a large engineering firm is typical of comments in just about every client survey:

> "I appreciate this conversation. It makes me think that [the sponsoring firm] really does see that we are all here to help each other, and I value that approach."

A Note About Perception

What should a firm do when it learns that certain perceptions that clients (or employees) have about the company are inaccurate? Although this is not an uncommon scenario, it's a difficult one for many managers to cope with, especially when the perceptions are negative. It is natural and extremely tempting to discount a perception that is deemed inaccurate; after all, that perception is *wrong*, isn't it?

In fact, that perception is not wrong at all. It is 100 percent true in the mind of the client (or employee) who holds the perception. And that is the truth with which strategic planners—and later, communications experts—must be concerned.

Rather than ignore these inaccurate perceptions as "outliers," a firm can instead view them as opportunities to refine its message in the marketplace and change or "correct" the perception. If a client study reveals a pattern of inaccurate perceptions, then the firm should develop approaches to change those perceptions in the marketplace as a whole or within the sector or sectors where the inaccurate perceptions seem to prevail. If it is an individual, or outlier, perception, then the firm should consider addressing it on a one-on-one basis with the client, for example by the project manager or principal-in-charge, as appropriate to the situation.

PRESENTING THE FINDINGS

With this quantitative and qualitative, internal and external information collected and analyzed, the firm can put together a management report that identifies the firm's strengths and weaknesses and focuses on the highest priority issues that it must address in its strategic plan. The management report can even—most often with a consultant's input—provide some preliminary recommendations regarding how best to address the key issues to provide food for thought for the strategic planning team before the planning retreat. The firm is now well prepared to develop the plan.

4

DEVELOPING THE STRATEGIC PLAN

"Dreams are but thoughts until their effects be tried."
—William Shakespeare

It takes time, effort, and commitment to gather the information your strategic planning team must analyze in order to identify the key issues that your firm will address in its strategic plan. It therefore follows that a firm should devote at least the same intensity of focus to developing the plan itself. You will need to determine who will participate directly in developing the strategic plan and then how you will structure the planning sessions, as well as the results you will challenge the planning team to achieve in those sessions.

In order to provide the opportunity to devote the attention that will be required, the strategic planning sessions must be unfettered by the typical distractions that we all face during the course of a workday. Most firms find it best to devote at least two full days to a facilitated strategic planning retreat in a location away from the office.

DESIGNING YOUR FIRM

Architects, engineers, and planners are engaged by clients to design a new physical reality. Because those of us in the design professions understand the design and construction process, we can compare strategic planning to designing a building. To relate the metaphor to something that many of us have experienced, consider the steps— based on the traditional phase-by-phase architectural/engineering design process—that an architect would take if you hired him or her to design a new house for you and your family:

Programming. At the outset of the assignment, the architect would visit you at home and interview you and your family to determine your functional and aesthetic needs and desires, and to see what you like and don't like in your current house. How many bedrooms and bathrooms do you need? Would you like a larger kitchen? Do you want a fireplace in your living room or family room? The list goes on.

This first activity as a prelude to design relates to the first step in the strategic planning process: conducting an in-depth and introspective analysis of a firm to identify what works and what doesn't work within the firm so that the strategic plan can preserve and build around its strengths and objectively identify the key **issues** that will affect its ability to succeed.

Schematic design. Next, the architect gets creative in order to translate your desires into a design represented by tangible images in the form of renderings, conceptual floor plans, and elevations, and your house begins to take shape on paper (or, more accurately, on a screen). When the architect has completed the conceptual design, he or she presents this depiction of the "finished product" to you to get you excited about your future in your new house and to obtain your buy-in for its design.

This design step relates to the very creative, big-picture process of developing a firm's long-term **vision** for its future in five or more years, describing the kind of firm you want it to be. At this phase in strategic planning, a firm will also develop a **mission statement**, an expression of its purpose and why it is in business.

Design development. In order to make the conceptual design more detailed and understandable so that your new house can eventually be realized, the architect refines the plans and elevations, describing basic building systems and materials and more actively involving other specialists such as mechanical, electrical, and structural engineers.

This phase in the design process parallels developing **strategies** to undertake and setting measurable, quantifiable **goals** to achieve in the midterm in order to move toward a firm's vision.

Construction documents. The house cannot be built—at least not easily or accurately—from design development documents, so detailed construction documents are needed to tell the contractor exactly how to build the house. These documents lay out unique conditions, giving the contractor a set of instructions from which he or she can convert the two-dimensional concepts into three-dimensional reality and achieve your vision in your finished home.

This final, most detailed design step, which lays out the project specifics, is parallel to developing **action plans** that serve to make the good ideas that are strategies tangible and "actionable." These nuts-and-bolts action plans will be just as detailed as a set of construction documents, with each task assigned to a responsible manager with an agreed-upon deadline for completion.

Construction administration. For better or worse, the architect knows that your house isn't complete when the construction documents go out the door. The architect still needs to monitor the construction to ensure that the project is being built according to the construction documents, address changing field conditions, answer contractor requests for information, adapt to late changes from you, and ultimately ensure that the built project fulfills the original design intent. The architect holds meetings and assesses progress regularly to make sure that everything is being done as prescribed, or that if a change is necessary in order to proceed, he or she makes it.

In strategic planning, you will also continue to **manage** your firm's progress toward its goals by meeting regularly with the strategic planning team, assessing progress toward completing action items, adjusting your course when necessary, and letting everyone in your firm know that you are making deliberate forward motion toward your firm's vision.

The elements of designing a project and strategic planning line up quite neatly:

Designing a Project	Strategic Planning

Programming
  • Determining the client's needs
	Issues
	  • What works and what doesn't work

Schematic Design
  • Creative, big-picture concepts
	Mission (permanent)
	  • Unchanging purpose of the firm
	
	Vision (long-term)
	  • A firm's aspirations for the future

Design Development
  • More defined building systems
	Goals (midterm)
	  • Quantifiable, measurable targets
	
	Strategies (midterm)
	  • Ideas to push the firm in the right direction

Construction Documents
  • Detailed instructions showing how to build
	Action Plans (short-term)
	  • Who is going to do what, and by when?

Construction Administration
  • Assessing and maintaining progress
	Communication and Implementation
	  • Sharing the plan and keeping it on track

It is important to understand that this strategic planning process—identifying key issues, creating a vision for the future of a firm, setting measurable goals, and developing strategies; and finally putting together specific action plans—will work well for any firm. However, as standardized as the *process* appears, it will yield a completely different plan for each firm. No matter what the project, every architect goes through the sequential phases of design that we described above—starting with big concepts and working down to the small details without missing a step or taking them out of order—and yet every design project is unique. Strategic planning employs a similar sequential process, yet because every firm's vision is unique—as are its issues or problems—every firm's strategic plan will also be entirely its own.

THE STRATEGIC PLANNING TEAM

Although it may not be the toughest decision that a firm leader makes in the strategic planning process, choosing whom to include on the strategic planning team should nevertheless be approached with consideration and care. The objective is to bring together strategic thinkers whose talents are both a complement and a counterpoint to those of the others on the team. These should be people who are committed to the firm and hold positions responsible for its success and who together form a group of a manageable size for making decisions. The team should be large enough to represent lots of different ideas that will inspire creative brainstorming and lively debate, and small enough to retain a sense of intimate teamwork and foster efficient decision making. In our experience, this translates to a planning team of from 5 to 15 members, with 8 to 12 being the "sweet spot."

The team should include those in your firm who fulfill the following functions, whether or not they have these specific titles:

- The firm leader: president, CEO, or managing partner/principal
- The chief operating officer (COO) or director of operations, if your firm is large or complex enough to have one
- The "business unit" leaders responsible for bringing in and executing the work in their target markets (these might be market leaders, studio leaders, department heads, or office managers); in some firms, these positions are defined by the firm's profit centers.
- The marketing and/or business development director
- The chief financial officer (CFO) or financial manager
- The human resources director
- Other key players or up-and-comers whom you would handpick because of their unique creativity, energy, and business intelligence that places them on your short list of future leaders of the firm

Identifying Your Strategic Thinkers

When you're thinking about whom to include on your strategic planning team in addition to the obvious senior managers, look around your
firm for other strategic thinkers. Perhaps seek a more junior person who is
smart, energetic, articulate, creative, and ambitious. This is someone who is
known for asking, "What if we did it this way, instead?" In fact, this person
may remind you of *you* when you were his or her age. They may not be
stockholders or key managers (yet), but these up-and-comers will add fresh
ideas, as well as a dose of realism, to your plan.

It also bears considering whom to *exclude* from the strategic planning team. Your firm may employ principals whose greatest value is in
some capacity other than as the firm's business leaders (lead designers
or technical experts, for example), and therefore the strategic planning
retreat is not a good use of their time. Or perhaps there are people who
have previously participated in retreats but no longer need to because
of their minimal contributions in past sessions. You must be frank with
yourself about who the noncontributors have been among your key personnel, and assign them roles other than on the strategic planning team.
The planning retreat must include real leaders and people with creative
and progressive ideas that they are willing to share.

The retreat participants (the strategic planning team) will have their
own views of the firm based on their positions. A branch office manager,
a marketing director, the lead designer, and the CFO could have very different perceptions and ideas about the firm, each centered around their
own priorities. It is very important in a retreat that everyone take a high-
level view of the entire firm, basing their planning decisions on the premise that what is good for the *entire* firm is good for everyone *in* the firm.

"MANAGEMENT" VERSUS "OWNERSHIP"

In making the admittedly difficult decision about whom to include
in—and exclude from—the strategic planning retreat, firm leaders
sometimes take the easy way out by simply including all of the firm's

shareholders or all those above a certain ownership threshold. This can be a mistake, because it fails to recognize the important distinction between the management of the firm and its ownership. As we discuss in Chapter 17, shareholders—the firm's owners—are essentially investors in the firm who vote for the firm's board of directors, which in turn appoints a president/CEO. Although they have the most direct financial stake in the firm, they do not necessarily warrant a role in its management or especially its strategic planning simply by virtue of their ownership share. The strategic planning team can just as likely include nonshareholders as well as owners—and neither based on their ownership status alone, but rather on their responsibilities within the firm and their potential to be valuable contributors to the strategic planning process and the firm's future.

T o o   M a n y   P e o p l e   t o   I n v i t e ?

Larger firms, in particular, often struggle with limiting the size of the retreat group. A 700-person E/A firm employed an interesting approach: the CEO decided to conduct separate sequential strategic planning retreats with two different planning teams: first, a "vision" retreat with a group of the most creative people in the firm regardless of their place in the management hierarchy (in fact, a significant number of partners were omitted from this group); and second, an "implementation" retreat to deal with strategies and action items, with primarily partners and other top management whose senior-level clout would be needed to implement what became the firm's sweeping reorganization.

THE STRATEGIC PLANNING RETREAT

In our experience, an off-site retreat provides the environment most conducive to developing a firm's strategic plan. A firm puts a lot of preparation into gathering information for strategic planning, and therefore should also go through a considerable amount of trouble to select exactly the right participants for the strategic planning team. The retreat forum gives that team the best opportunity to focus on developing a strategic plan that addresses the key issues identified in the information-gathering phase.

Retreat Location

To some, the word *retreat* conjures images of golf outings and tiki bars, palm trees and karaoke with comrades enjoying some well-earned R&R together. This sort of team-building getaway may be appropriate— in firms that can afford it—as a reward for achieving certain financial goals or to wind up an annual shareholders' meeting, but not for the intellectually demanding work of strategic planning.

It may seem unimportant or even frivolous to expend company resources and key people's time to take part in an off-site retreat, but for strategic planning, the isolation from daily routines is necessary. Strategic planning retreats are intense working sessions. They are not mini-vacations on the company dime, as anyone who has participated in a successful strategic planning retreat will attest. Therefore, a firm should choose its off-site retreat location carefully.

Most hotels have well-equipped facilities to house a strategic planning retreat. The operative word is *function*, not *luxury* (although a few creature comforts don't hurt, either, as long as they are not so excessive that they appear to be an embarrassment of riches for the retreat participants). A checklist for an ideal retreat location includes the following:

- A meeting place away from the office, where planners will be free from phone calls, e-mails, interruptions, and other distractions. It should be far enough away to discourage impromptu "visits" to or from the office, but reasonably accessible to all participants, so that meetings can begin on time and progress without worries about an interminable drive home after a hard day. When a firm decides that its retreat should be in an out-of-town venue, then everyone stays overnight in the same facility, and commuting distance is not an issue.
- A comfortable space that is equipped to support planning activities, such as a hotel or conference center meeting room. If planning team members will be visiting from far away, it is best to hold the retreat in a hotel where they can also have rooms for the night, and that is also easily accessible to the airport, train station, or major highway interchange, as appropriate.
- Enough "elbow room" around a big table so that everyone can spread out their materials while still being able to make eye contact with their colleagues and participate actively in discussions.

- A facility where meals and energy-sustaining snacks will be provided. Plan to provide a real breakfast, a balanced lunch, and food and drink during breaks that will help to keep up planners' energy levels throughout the day.
- Windows and plenty of natural light. As lively as strategic planning sessions can be, a windowless room becomes oppressive to even the most enthusiastic participants and can put a damper on the entire process.

Retreat Process

It might surprise you to learn that the most effective strategic plans—for firms large or small—should be just 8 to 12 pages in length. That's it. More bulk and too much detail only detract from the plan's serving as a user-friendly management tool. (This is in contrast to the management report and employee or client survey reports, which contain a great deal of information and can therefore become voluminous documents.) Developing such a streamlined plan from the myriad data and ideas that a firm collects is no easy task, however. That is why a retreat facilitated by an expert consultant is beneficial—if not vital—to developing the strategic plan.

Retreat agenda. The retreat should start out with introductory remarks from the firm leader to set the tone for the entire event. That should be followed by the facilitator laying out some basic ground rules for participation, some of which are procedural ("everybody turn off your cell phones"), while other, more substantive items put the participants in the proper frame of mind to begin developing a plan for the firm. Then, encouraging all the participants to describe their own expectations for the retreat and what they want to achieve gets everyone used to talking and participating, as well as ensures that the retreat will accomplish what the group wants it to accomplish.

Prior to the retreat, the planning team members should have reviewed the management report and, if applicable, employee or client survey results and noted any questions or comments. Discussing people's reactions to the reports serves as a bridge between these "opening ceremonies" and the actual planning and allows any pent-up thoughts and concerns to surface.

When engaging in the actual strategic planning (which occupies the remainder of the session), the group works from the big picture down to the details. An inspirational and appropriate first planning activity is to develop the firm's mission and vision, or to update the vision if the firm already has one. See Chapter 7 for a detailed discussion of how to craft a mission statement; Chapter 8 addresses a firm's vision.

The retreat should then continue with a review of the key issues identified in the management report. It is important that the entire group validate the key issues in direct and unambiguous language that everyone in the planning group and also in the firm can understand. Often, firms have more issues than a single plan—or more accurately, the members of the planning group—can address and implement in the first year. It is important for the group to be able to prioritize the issues, picking and choosing those that the group can "afford" to do (not just in terms of money) and making sure that it tackles those that the firm cannot afford to ignore.

Once the group has developed or updated the firm's vision, the bulk of the working sessions within the retreat will consist of developing goals, strategies, and action plans to address the key issues in order to reach the vision. It is important to limit the key issues to be addressed in the strategic plan to a reasonable number—usually no more than five. This may involve rigorously prioritizing the top issues from a longer list and reserving the others for a future strategic plan update. See Chapter 9 for further advice about identifying and addressing key issues. Chapters 10 and 11 lay out how to develop strategies to address key issues and achieve goals, and Chapter 12 explains how to craft action plans.

Retreat protocols. In keeping with the "no interruptions" theme for the retreat, planning team members should turn off their cell phones, PDAs, and anything else that might buzz, ring, or vibrate on the tabletop. One firm has a "cell phone basket" into which everyone must place their phones upon entering the meeting room. Another imposes a strict $50 fine—payable immediately in cash—on anyone whose cell phone rings during a retreat. Because retreats are most often on workdays, it is important to build midmorning, lunchtime, and midafternoon breaks into each day's retreat agenda that allow 10 to 15 minutes for team members to check their messages and schedule return calls for a time following the retreat.

Although the plan itself should be shared—at the appropriate level of detail to be determined after it is completed—with the firm's staff, it is crucial that the background discussions that go into the plan stay in the retreat room and remain confidential within the planning group. On occasion, a firm leader will hesitate to discuss certain items with the planning group for fear of "leaks," which is an immediate red flag signaling that there may be people in this select group who lack the maturity to keep things confidential, and therefore shouldn't be included on the strategic planning team.

Within the walls of the retreat room, there are to be no "sacred cows." Even the most sensitive topics are up for discussion. Just because a firm has always done something a certain way doesn't mean it has to keep doing it that way. Just because a firm has never done something doesn't mean it can't examine the pros and cons of trying it. In fact, maintaining the status quo is the antithesis of a strategic plan because a plan, by definition, is about changing a firm for the better.

If you've brought the right players on to the strategic planning team, you can be assured that the retreat will involve plenty of discussion and debate. We encourage healthy debate for a number of reasons. First, it allows team members to air issues and express their opinions. More importantly, this kind of debate allows one "side" (or individual) the opportunity to convince the others of the wisdom of a particular position. Together, the group engages in these discussions and decides how to move forward to the next issue. All of the strategic planning team members should agree to some ground rules for the discussion and decision-making process.

Discussion of a particular topic should end when everyone in the group agrees to go along with a collective decision, whether or not that decision was their first—or even their second—choice. When the group makes a decision, all of the members agree to follow the firm leader's example to conspicuously support it, without exception.

There may be times when a team member sits back, often with arms crossed, and refrains from participating in a discussion. It's possible that the team member is simply listening and learning rather than contributing to the discussion. However, if the reticent behavior continues, it may be a sign that this person cannot or will not agree with the larger group's decisions and may harbor ideas that he or she will keep to himself or herself or—far worse—share outside the meeting with others. This behavior undermines the group's effectiveness in formulating strategies and action plans that everyone will support and communicating a unified message to the firm as a whole.

On rare occasions, a planning participant may decide that the firm is heading in a direction with which he or she disagrees or that is incompatible with his or her professional goals. If this leads to a parting of the ways—as it may—that is a more productive outcome for both the individual and the firm than having an undercurrent of "no confidence" emanating from this erstwhile key person, which will always be perceived at some level by the rest of the staff.

On Building Consensus

Consensus building is a term bandied about a lot these days. The term is used somewhat loosely to mean getting a majority of stakeholders to buy into a decision or course of action, or sometimes even getting the naysayers to accept a course of action with which they disagree. But strategic planning groups should almost never have a binding vote to make a decision. When a group votes under the familiar terms of "majority rules," the decision takes on the aura of a competition with winners and losers. Those in the "minority" may feel like losers and be unlikely to support the decision.

Let's be very clear about what we mean when we say "consensus." It rarely connotes unanimity, because, no matter how long they have worked together, any group of people will have differing ideas on any given subject. In a strategic planning retreat, the goal is that everyone in the room—the entire strategic planning team—affirms that they will actively support a decision that the group makes at the end of a debate. No matter how they felt about the topic at the beginning of the discussion, every member of the group *must* align and present a unified front to the rest of the firm (and the community at large). When the group reaches consensus, no one outside the room should be able to tell who was initially in favor of the decision and who wasn't.

To emphasize the point: consensus isn't "agreeing to disagree." It's the opposite: agreeing to *agree*, publicly, conspicuously, and enthusiastically. This is the consensus that is necessary to develop a strategic plan that engenders firm-wide buy-in and can succeed.

Retreat facilitation. Strategic planning retreats work best when they are facilitated professionally. The facilitator—optimally a strategic planning consultant—keeps the group focused on mission, vision, issues, goals, strategies, and action plans, and also—because he or she understands the industry—knows when and how to ask the kinds of provocative questions that will keep the group thinking strategically. The facilitator will also give the group a reality check, noting that the firm may be embarking on a path that has caused problems for other design firms.

The facilitator will ensure that the group addresses all of the agenda items necessary for a complete and practical strategic plan, and that for every issue, the group decides on at least one strategy (but quite often, several) to achieve it. In turn, the facilitator will assist the group in brainstorming action plans to carry out in pursuit of each of the strategies.

As the retreat progresses, the facilitator will document the discussions and decisions that the strategic planning team makes and ultimately distill the mission, vision, goals, strategies, and action items into a strategic plan that is concise, clear, manageable, and measurable. The facilitator will also ensure that each action item is assigned to a person responsible for seeing it through by a specific due date.

Once you have completed the big job of creating a strategic plan, you're entitled to a sense of satisfaction and accomplishment. However, the retreat should conclude with one more very important activity: discussing how to communicate the plan to the rest of the staff, which is really the first step in implementing the plan.

5

BRINGING THE PLAN TO LIFE

"A good plan executed today is better than a perfect plan executed at some indefinite point in the future."

—General George Patton

A good strategic plan has credibility; all of your firm's employees should be able to buy into it. It should be bold and direct enough that the staff will proudly hold themselves and one another accountable for sticking to it. You probably developed the plan at a very high level in the firm, with key leaders, managers, and recognized talents participating on your planning team. The air may indeed have felt rarefied in the retreat room, and the enthusiasm likely built to a crescendo as the group approached completing the firm's road map to the future.

However, it is important to remember that implementing the strategic plan should be every firm member's business and responsibility. Everyone in the firm—and we do mean everyone—can and should play a role in the plan's success; their knowledge and involvement is the first step in bringing your strategic plan to life.

SHARING THE PLAN

To conclude the strategic planning retreat, the planning team should discuss how best to share the plan with the firm's staff. Strategic plans are far more successful when everyone in the firm understands and buys into them, and that requires a certain amount of disclosure on the part of the planning team and, even more so, the firm leader. Following a few simple guidelines will allow staff members who were not in the planning group to feel like participants in the process and contributors to the results:

- **Promptness**. Communicate with the staff immediately following the retreat. In small firms, get everyone together for a quick debriefing; in larger firms, the firm leader should send an e-mail to all employees letting them know that the strategic plan has been developed in the retreat by the planning team and summarizing the key issues very generally. Don't keep people waiting or guessing. Everyone knows that the planning group is meeting in seclusion, and promptness will preempt the churning of the rumor mill.
- **Candor**. Talk openly with everyone in the firm. Many firm leaders feel unnecessarily reluctant to discuss firm business with the staff. Be enthusiastic when sharing the future vision for the firm and be candid when discussing what it will take to reach it. When in doubt, err on the side of straightforward disclosure. Granted, there may be elements of the plan that are not suitable for mass consumption, but those sensitive elements are rarely the primary components of the plan and are hardly an excuse not to share the plan broadly.
- **Publicity**. Hold an all-hands staff meeting (or series of meetings, if your firm is large or has multiple offices) to explain the planning process and the plan itself, and to stimulate firm-wide support. If your firm is large or geographically dispersed, post the plan on your intranet where people can find it for reference. Leaders of firms with multiple offices often put on a "road show" in which they personally visit each office to present the firm's plan and answer questions face to face.
- **Tangibility**. Publish a version of the strategic plan for all employees. Give the staff something to hold, and even to take home. People need to feel trusted, and if they can be trusted to design

a bridge, building, or water treatment plant, it is insulting not to trust them to know the future of the firm to which they contribute their labors. Considering the long hours many design firm employees work, they also need to be supported on a personal level by their families, and bringing home a printed version of their firm's strategic plan facilitates that level of support. Many firms develop attractive graphic pieces to give their employees, ranging from a simple laminated placard displaying the firm's mission to trifold brochures containing the essence of the entire strategic plan, to mini-CDs containing the firm leader's slide presentation of the plan.

- **Reporting Progress**. As your firm implements its action items and achieves its goals, keep the entire firm apprised of the progress. The firm leader should reassure everyone about the plan and the direction the firm is taking by personally meeting periodically with groups of staff, posting updates and milestones achieved on the company intranet, and reporting successes—with kudos to people by name—through regular e-mails and/or in an in-house newsletter.

If You're Reluctant to Communicate

As important as candor is, there are still firm leaders who, out of some misguided fear of industrial espionage among design firms or a lack of confidence in the interest and maturity of their employees, display alarming reluctance to discuss the strategic plan with their full staff. This objection is difficult to understand. If, in fact, a job candidate came to an interview at your firm brandishing the strategic plan of his or her current (or former) employer, how seriously would you consider hiring someone displaying those ethics?

MANAGING CHANGE

Strategic plans always connote change for a firm. However, when given a choice, people often instinctively opt for the status quo and do not readily adapt to change. Here are some lessons learned from companies who have successfully smoothed the path for change:

- People need to understand the reasons for any change. A thoughtful plan helps everyone understand and buy into change; an arbitrary edict from the CEO—"an idea I had driving to work this morning"—seems spur-of-the-moment and causes people to resist change.
- People need to see how they will benefit from change. Firms that share information freely have staffs that embrace changes rather than deflect them.
- Longtime staffers may be more defensive about changes than newer employees, perceiving anything that rocks the boat as a threat to their stability and security. This further points to the need for more and better communication from the CEO about the benefits of change.
- The need for change is most apparent and easiest to understand when it is about solving problems rather than taking advantage of opportunities. Firms with serious troubles understand that they have to change, sometimes in order to survive. It is more difficult to convince staff of the need for change when a firm is already successful, yet has even higher ambitions.

The lessons about managing change are that it is all about straightforward communication, sincere inspiration, and bringing everyone in the firm into the planning process in one way or another, encouraging them to feel that the firm's strategic plan can also serve as their own career path.

IMPLEMENTING THE PLAN

Accountability is the watchword in implementing your firm's strategic plan. From the very top—the firm leader—the message must permeate throughout your firm: "We have a great plan, and we are following it!"

The firm leader must be a champion for the firm and the plan, and simultaneously hold people accountable for implementing their action items within the plan.

In order to keep the plan fresh and alive, the strategic planning team should stay together, rather than disband upon publication of the plan. The team as a whole—or at a minimum those who have responsibility for action items—should hold regular monthly meetings to review activities undertaken in pursuit of the plan and to share progress toward achieving its goals. Keep the meetings concise—no more than one hour, and by no means a "mini-retreat" with new debate about plan elements—by using the overall action item list as the meeting agenda. And *always* report progress to the entire firm on a regular basis.

The Leadership Factor

The leader of a firm—whether titled president, CEO, managing principal, or otherwise—is the "project manager" of the strategic plan. In fact, the job description of a CEO could be summarized in the phrase "to develop and implement the firm's strategic plan." Just as land planners and real estate developers analyze a piece of property for its "highest and best use," the same is true when prioritizing the responsibilities of a firm leader. Of course, in smaller firms, the leader has to wear many hats. Nonetheless, in any firm the leadership to carry out the actions detailed in the plan, as well as the mandate that those responsible for implementing them are held accountable for their progress, must come from the very top.

A common pitfall for design firms in strategic planning is the tendency to overplan and underimplement. In a planning retreat with creative people, ideas come readily and words on flipcharts flow easily, but back in the office, following through on actions can be difficult. To avoid this, the firm leader must leverage his or her influence to motivate everyone in the firm with enthusiasm for the strategic plan and hold people accountable for carrying it out.

Personal Accountability

The new president of one civil engineering and surveying firm with four offices assumed personal responsibility for a number of action items in his firm's new strategic plan. As part of publishing the plan, he took the list of action items, including his own (which were many), enlarged it into a poster, and mounted it on the walls in each office's break room next to the soda machine. In this way, everyone in the firm was aware of who was responsible for which action items; it was impossible to buy a soda without being reminded of the status of those responsibilities. Then, as each action item was completed—again, including his own—the president visited each office, conducted a brief staff meeting, and ceremoniously put a big red check mark next to the item on the poster.

KEEPING THE PLAN UP TO DATE

The world is not a static place; things change constantly outside in the marketplace and within the walls of your firm. The economy rises and falls, competitors move in and move out, and clients—and employees—come and go. Therefore, strategic planning has to be thought of as an ongoing activity, not just a one-time or even an annual event. The plan is a dynamic tool that your firm should continuously revise and update as necessary to respond to internal and external forces that change the circumstances of your firm or market. Firms typically make a lot of progress during the first year of their strategic plan because of momentum carried over from the group's enthusiasm during the development of the plan. However, making the plan happen places a significant burden on the key players who are in charge of implementing the action items, most of whom—because of their management standing in the firm—have responsibilities that also include client relationships and project performance.

Of course, a firm can expect some low notes as it embarks on implementing the strategic plan. If it becomes apparent that any of the key players are not able to shoulder their load, the firm leader should reassign their responsibilities rather than let their tasks wane. After all, getting items accomplished is more important to the firm than who accomplishes them. In addition, a firm can expect some of its old issues, such as clearing out nonperformers, to linger even after they are addressed in the strategic plan, simply because managers often avoid confronting them directly.

Therefore, it's a good idea to have a checkup at the first anniversary of the strategic plan. Many firms call their strategic planning consultant at this time to conduct short interviews with the firm's management to help identify new issues that have arisen and assess which, if any, of the previously noted issues are languishing unresolved. In this way, the consultant can serve as the "conscience" of the firm, especially when it comes to dealing with particularly difficult issues. The one-year mark is also an opportune time to conduct a firm-wide employee survey to gauge how the entire firm feels about progress and unearth any new issues.

Each year, firms should conduct a one- to two-day facilitated retreat to review and possibly update—but not overhaul—the firm's long-term vision, identify new or lingering issues, and develop new strategies and action plans to address those issues to clear them from the path of success once and for all. In the end, it is the firm leader's job to institutionalize the pursuit of the firm's vision and goals—through the strategic plan—into the daily life of the firm.

6

THE FIRM LEADER'S ROLE

"Leadership is the capacity to translate vision into reality."
—Warren G. Bennis

In a design firm, there may be a number of partners or accomplished individuals who have risen to positions of leadership (and possibly ownership). In many—if not most—firms, there is a single firm leader whose broad focus is on the future of the firm and who is acknowledged to have the judgment, discretion, and recognized authority to make the best short-term decisions toward the firm's long-term vision. Even if there are many partners or owners in a firm, the experiences of successful firms show that there must be a single leader in a position to advance the interests of the owners and of the firm as a whole.

When we use the term *firm leader,* we refer to the individual who may have one or more of any number of titles—depending on the firm's business or corporate structure, or sometimes simply tradition—such as chief executive officer (CEO), president, managing partner, managing principal, or principal, and who is directly accountable to the firm's board of directors, who represent the interests of the shareholders. In many design firms, especially smaller firms or those with concentrated ownership, a few individuals may occupy multiple positions, each serving several roles. Whatever the leader's title, all employees are either directly or indirectly accountable to this individual for their performance. (For a detailed discussion of firm management, see Chapter 15.)

THE FIRM LEADER AND THE STRATEGIC PLAN

The firm leader's active participation in—and commitment to—the development and implementation of the strategic plan is vital to the plan's success. If the leader does not wholeheartedly believe in the power of strategic planning and support it, the odds are that the planning process will fail. Only the leader is charged with maintaining the necessary focus on the planned future of the firm and has the authority to mobilize the entire staff to achieve it.

There are many dimensions to the firm leader's responsibilities, but stated succinctly, the leader's job regarding strategic planning is to help others share a vision of where the firm is going and determine how to get there. Successful businesses tend to have a singular leader, someone who articulates the firm's vision and inspires the entire organization around it. The leader carries the torch for the firm, lighting the way to its future. This is true whether or not a firm has a strategic plan, but without the road map that a plan provides, the task is far more difficult; the leader is forced to improvise as things go along. It would take an exceptionally strong and capable leader to create a singular vision for a firm in the absence of a strategic plan, then articulate that vision clearly to the staff, enlist their support for it, and ultimately drive the entire firm toward achieving it—in fact, we can't think of a successful design firm whose leader operates effectively in this continuously improvisational way.

Another way to summarize the firm leader's job in the context of strategic planning is to lead the development and implementation of the firm's strategic plan, and to propel the firm toward reaching the vision described in it. The leader needs to have intimate knowledge and awareness of what is happening inside the firm and outside in the marketplace and be able to discern on a day-to-day, month-to-month, and year-to-year basis where the company should be going and what he or she needs to do to help it get there. The strategic plan's vision, goals, and strategies provide the valuable perspective and context for the decisions that the leader and other managers have to make every day.

The strategic planning process itself can help a firm leader maintain long-term focus and keep his or her firm leadership role foremost in mind. As discussed in Chapter 4, the process begins with the big picture of defining the firm's mission (see Chapter 7 for more about mission) and describing a destination in the form of a vision (see Chapter 8), and culminates in a detailed set of action plans (see Chapter 12) that serve as a turn-by-turn road map for implementation. As a firm's strategic planning team develops the plan, they assign action items to individual members of the team or sometimes to other members of the firm. Certain action items are appropriate for only the firm leader to carry out, because they involve his or her broad knowledge of the firm or require some element of firm-wide authority or influence; they simply should not be taken on by anyone else. There are, however, many action items that can and should be performed by members of the firm other than the leader. Assigning action items to others, as appropriate, helps the firm leader resist the temptation to micromanage and also strikes a balance between taking on an unrealistic amount of responsibility and pushing too much responsibility away.

The firm leader should work with the strategic planning team to determine the necessary action items and who is the best person to be responsible for completing each one. In this way, the strategic planning process also helps to reinforce the firm's culture and build collaboration among the members of the management team.

STRATEGIC PLANNING CHALLENGES FOR FIRM LEADERS

With all the benefits that the strategic planning process affords, it can also pose two distinct challenges to a firm leader: the first relates to the unknowns that the process can reveal, and the second has to do with the planning retreat itself.

A potential difficulty for the firm leader is that once the strategic planning process has been set in motion and has begun to unearth the key issues affecting the firm, some of those issues may relate to the leader's own foibles. This would be an uncomfortable position for anyone, but especially so for a firm leader. Once senior managers have been interviewed and the firm has gathered additional information in preparation for strategic planning, the firm leader may feel that he or she is losing control of the outcome of the process. The leader may feel uneasy about someone—often a consultant—asking probing questions and uncertain about the results of the process. The level of anxiety that such questions cause the leader, however, may be a reflection of the validity of those questions. It takes courage, maturity, security, and a willingness to place the greater good of the whole firm ahead of one's own ego to confront these issues with honesty, humility, and grace, as well as with authority.

The strategic planning retreat is an intense, one-, two-, or even three-day working experience for its participants. With appropriate preparation, the retreat should be productive and successful, and the strategic planning team will make many decisions with substantial long-term impact for the firm in that concentrated period of time. This can actually be the source of considerable and unexpected anxiety for the firm leader, particularly one who has been in charge of the firm for a very long time (often, though not always, the firm's founder) and who is used to unilateral—and even autocratic—decision making.

A Tale of Two Leaders

The role that the firm leader takes in strategic planning is a driver of success, as evidenced by these examples:

- The CEO of a large architecture firm recognizes that every few years, the firm needs to develop an updated and newly relevant strategic plan, and that an important aspect of his job is to shepherd it through and roll it out to the staff. At the same time, this CEO knows that developing an updated vision for the firm requires buy-in from other people, and that the implementation of the strategic plan depends on the support and work of many others. This CEO "gets it," and as a consequence, his very successful firm gets behind the strategic plan.
- The leader of a 50-person design firm commissions a strategic plan every three or four years, but only reluctantly, and only under

pressure from the next generation of firm leaders. She tends to concede the need for a plan in response to a crisis situation, or in response to increasing grumbling from subordinates. Her firm fails to derive the full benefit of its strategic plans and actually suffers a greater loss of morale due to raised but unmet expectations. The firm leader's attitude also reflects a major issue for the firm: her continued reluctance to take the initiative in leading the firm through the planning process calls into question her overall leadership effectiveness.

DECISION MAKING DURING THE PLANNING RETREAT

The strategic planning process is designed to encourage people to speak out and put forth their ideas and opinions about the future of the firm, the strengths that will move it forward, and the obstacles that stand in the way of its success. Although many decisions are made in the course of the retreat itself, some decisions are simply too complex to be made in a concentrated period of time. The planning team should address this reality by creating action items that allow the firm leader and others to make better informed and well-founded decisions at a later date.

For example, a firm with multiple offices or markets, or one that provides a variety of services, may wish to focus its energy and resources on areas of the firm that are performing the best, and cut back on or even shut down a business component that isn't performing. But even though the participants in a strategic planning retreat recognize the need for action, they may not have sufficient information to make the correct determination within the confines of the retreat. The strategic planning team and the firm leader may know that one or two offices must be closed, or that the firm may need to enter a new market, but it may be impossible—due to a lack of immediately available data—to decide in the planning retreat to "close the Dubuque office" or "get into the higher education market." Instead, the group might define as one of the firm's strategies focusing on its highest-performing markets through its best offices and cutting its losses where the firm has historically underperformed. The group could then define three action items: (1) to conduct a financial analysis of the performance of each of the firm's business units; (2) to research which markets offer the best short- and long-term prospects; and (3) to charge the firm leader (independently or in concert with other senior managers) to make decisions to implement the defined strategy.

With proper preretreat preparation, many decisions, even big ones, may become patently obvious during the retreat, but some won't. Decisions that *can* be made during a retreat *should* be made during the retreat. Deferring too many decisions results in a strategic plan that doesn't define any direction or make any determinations at all, diluting the impact of the plan and potentially defeating its whole purpose. The retreat participants of a transportation engineering firm, for example, might know enough about a certain state's department of transportation market to make the decision on the spot during the retreat to exit that market because there is too much local or parochial competition with preferential status, the state procurement process is fee-based rather than qualifications-based, or the volume of available work is too small and too far away from the firm's nearest office to be profitable.

THE RELUCTANT LEADER

Not all firm leaders embrace the idea of strategic planning for their firms. When it comes to strategic planning, it is nearly impossible for subordinates—even highly placed managers—to push a reluctant firm leader from behind with any hope of success. There is an old saying, "You can't push rope." The firm leader must be the true strategic planning champion, if not initiating the concept, then embracing it and maintaining its momentum year after year. A strategic plan that is not supported—or worse yet, is casually undermined—by the leader may do a firm more harm than good.

Strategic planning can be a cathartic experience in that it has the ability to expose a firm's most sensitive issues, which are often well known to all, but up until the time of the planning process may not have been addressed. This situation is particularly acute when the issues revolve around the firm's leader. When this happens, the planning process raises the expectations of the participants and the staff as a whole that the issues will (finally) be addressed. However, if the leader is unable to recognize the validity of the issues or unwilling to address them constructively, there is little prospect of meeting those expectations, which results in deeper disappointment than if an attempt at developing the strategic plan had never been undertaken. All that is accomplished is that people's expectations have been raised, only to be dropped from a greater height.

One Reluctant Leader

The outgoing founder of a 100-person A/E firm had planned his retirement and named his successor as the new president. Eager to place his imprint on the firm, the ambitious incoming president embarked on a strategic planning process and appeared to be doing everything right, including a thorough analysis of the firm's systems and plenty of interviews with the senior managers and key staff. The consultant prepared a management report, and a group of 12 managers—including the retiring founder and the new president—held a very productive two-day retreat at which everyone was an active contributor. At the conclusion of the high-energy retreat, with the walls of the conference room papered with the group's vision for the firm as well as their strategies and action plans, the soon-to-be-retiring founder stood up, thanked his key managers for their "thoughts," and announced that he would "take it all under advisement." The deflated looks on the faces of the participants, especially the new president, told the story about what was wrong with the firm's leadership in the past, and what they yearned for in the future.

COMPONENTS OF STRATEGIC PLANNING

7

YOUR MISSION: THE PURPOSE OF YOUR FIRM

"When you discover your mission, you will feel its demand. It will fill you with enthusiasm and a burning desire to get to work on it."
—W. Clement Stone

When we give seminars on strategic planning at professional conferences—the attendees of which are usually design firm principals—we often ask attendees to raise their hands if their firms have a mission statement. Typically, about half of the hands go up, sometimes haltingly. We then ask those who claim to have a mission statement whether they can recite it, and nearly all of the hands go down. To make it a little easier, we then ask how many could summarize or paraphrase their firm's mission statement. A few hands tentatively go back up. The point of the exercise is to demonstrate that a firm's mission statement must not be very meaningful if even the firm's principals can't recall it.

The term *mission statement* often causes eyes to roll and evokes visions of long-suffering comic-strip cubicle denizen Dilbert, and with good reason: many mission statements are nothing more than strings of meaningless jargon and management buzzwords. Employees' attitudes toward mission statements have become so jaded that there are even satirical websites that "compose" mission statements from combinations of "management speak" phrases. Despite the cynicism that many mission statements have engendered, an effective mission statement can be a powerful rallying point for your firm. Every organization has a mission, a purpose, a reason for existing—something that distinguishes it from other companies—whether or not it is stated. Often, that purpose was the gleam in the firm founder's eye from the very beginning. When you and every member of your firm understand the mission and can readily articulate it, it can be of tremendous value.

The Difference Between Your "Mission" and Your "Vision"

Don't confuse your firm's *mission* with its *vision*. Although the two terms sound similar, they are inherently different things with different purposes. The *mission* captures the essence of your firm, distilling it into a unique, concise, and memorable statement of why it is in business. The *vision*, which we discuss in Chapter 8, describes where your firm is going and what you want it to be at some point in the future.

STATING YOUR PURPOSE

For the most part, the elements of a firm's strategic plan are private and proprietary, your firm's own road map to its future. The mission, however, is an exception; the mission statement explains the purpose of your firm—the reason for its existence—and it is entirely appropriate for both internal and external consumption. Firms that take full advantage of the power of their mission statement emblazon it on their business cards and stationery, websites, e-mail signature lines, lobby plaques, coffee mugs, T-shirts, baseball caps, brochures, proposals, and many other visible media. When a firm has a memorable mission statement, promotes it continuously, and refers to it often, it gives everyone inside and outside of the firm a common understanding of the firm, and it rallies everyone under the same flag.

It is up to the leaders of the firm—the president, CEO, and principals—to *live* the mission. Firm leaders should be the walking, talking, living embodiment of the mission statement—role models who set the bar and exemplify the firm's mission for everyone else.

Your firm's mission should remain constant over time; because it articulates why a firm is in business, it should be permanent. (Of course, in the unlikely event that the mission of your firm should change drastically, so should your mission statement.) Furthermore, it would be difficult to memorialize your mission statement in the minds of your employees, clients, prospective clients, consultants, and others if you revise it frequently. Because of its importance in defining your firm's enduring corporate identity, the mission statement wording is worth the extra time it takes to polish. In every other element of a strategic plan, the content is much more important than the phrasing; but in the case of the mission statement, both the message *and* the way it is expressed are crucial to its effectiveness.

MISSION STATEMENTS THAT WORK

To be effective, a mission statement should be sincere, uplifting, and inspiring, expressing the core principles of your firm's corporate culture to help employees understand why they come to work every day and to explain to clients and prospective clients how the purpose of your firm fulfills their needs. It should be concise and easy to understand, and therefore easy to remember. After all, if you cannot remember it, you cannot live it. An excellent example from outside the design professions is Google's mission statement, "…to organize the world's information and make it universally accessible and useful."

For a mission statement to be truly inspiring and memorable, it must also be unique to your firm. Firms should take care to avoid certain overused words such as *quality, excellence,* and *service* in their mission statements. In the realm of professional services, quality, excellence, and service are the baseline; clients expect these attributes of *every* design firm and professional. There is nothing unique about them. Proudly proclaiming a basic level of what clients expect is no way to distinguish your firm from any other.

When your mission statement is a winner, people remember it, and it tells them something special about the firm. In our practice, we see scores of mission statements, good, great, and otherwise. One of our favorites is the mission of a 40-person architecture and planning firm that designs civic buildings such as public libraries, community centers, and public safety facilities:

Realizing our clients' visions and building community through the pursuit of the craft and beauty of architecture.

With key words such as *community, craft,* and *beauty,* this mission statement imparts the sense that this firm believes that community buildings should be programmatically functional as well as beautiful; this is the kind of firm that is meticulous and caring and wants to make your town a better place to live. Compare this mission statement to that of an ambitious 90-person architecture firm with a strong work ethic and dedication to its clients. Its mission, which for emphasis is always expressed as three sentences on three separate lines, is:

Create enduring value for our clients, their communities, and our firm.
Do what it takes every day.
Think BIG!

Now there's a firm with confidence! Their mission statement leads you to expect that this firm engenders a lot of loyalty among its employees, many of whom you would expect to have shirts or jackets with the firm's logo on them. Although at first glance the mission may appear to be "marketing language," the managers and employees of this firm truly believe it, so much so that "doing what it takes every day" is an important element in the firm's employee performance reviews. This firm is dedicated and "walks the talk."

A rapidly growing engineering firm of several hundred people with an aggressive culture and an attitude in tune with its entrepreneurial and fast-paced real estate developer clients is driven to be number one in the markets it serves, and it works diligently to achieve that goal. Appropriately, it has a very assertive firm mission statement:

We will dominate every market we serve through total commitment to our clients' success. We will be the best: every project, every day.

This is a certainly a focused and determined firm!

A mission statement in a completely different vein comes from a 90-person civil engineering, planning, and landscape architecture firm. The people in this firm really enjoy what they do, which is mostly large-scale planning work. They also fervently believe that what they do improves their communities, and they are unabashed about their desire to earn rewards from the results of their work. Their mission, which they also express in three sentences on three separate lines, perfectly sums up their reason for existing:

Have a little fun.
Make a little money.
Do some good.

Another effective mission statement comes from an architecture firm that designs hospitals. It cares deeply about the end users of its work: the patients. The firm's mission statement is the most pointed and specific of all these examples:

Promoting wellness through architecture.

These professionals really believe in the contribution they make. The day after this firm developed its mission statement in a strategic planning retreat, it immediately and proudly began to incorporate the statement into its stationery, website, and everything else that could help the employees communicate their mission to everyone they know—and people they would like to know as clients.

All five of these successful firms undoubtedly provide quality, excellence, and service to their clients. But their mission statements immediately distinguish them from other firms. Using very few words, these statements resonate with their current and prospective clients and motivate employees with a common purpose.

More on Missions

For more inspiration on crafting a mission statement, we recommend *Say It and Live It: The 50 Corporate Mission Statements That Hit the Mark* by Patricia Jones and Larry Kahaner. As its title suggests, this book is a compilation of mission statements for all kinds of companies, large and small. Each case presents an example of how the subject companies have been successfully guided by their missions.

Your firm's mission should serve as the centerpiece of its strategic plan, a touchstone in light of which management will consider all future decisions by asking the nonrhetorical questions: Is this consistent with our mission? If not, why would we do it? You will find that decisions that seem agonizingly difficult in the abstract are much easier to make within the context of a clearly defined mission in which everyone in the firm believes.

WHAT ABOUT VALUES?

Some firms place added emphasis on articulating their values—also known as core values—which generally describe the ethical rules or ideology by which the firm operates. These values are different from the mission or purpose of a firm, but are no less important to firm leaders who want to codify them.

There are three keys to making a statement of values work for your firm:

1. **Sincerity**. First and foremost, the leaders of your firm have to genuinely believe in the values. Otherwise, they will inspire cynicism from the start.
2. **Reinforcement**. The leaders have to preach and refer to the values constantly so that they permeate the thinking of your entire firm.
3. **Accountability**. Everyone in your firm has to agree to be held to the values. As soon as someone appears to be "getting away with something"—especially a manager—the sanctity of your firm's core values will begin to erode, and they will become worse than worthless.

A set of values that everyone understands and buys into can be important because they build a sense of trust—up and down the organization—that people will perform and behave in accordance with the values.

If you are ready to articulate your firm's core values, consider the following good examples of value statements from other design firms:

> **Integrity:** We expect honesty and integrity to occur without thought. Living this value will produce enduring relationships with coworkers, clients, and others.

Teamwork: We recognize that only through successful team-work can we achieve our mission and expand our boundaries. We believe that our strengths and potential lie not in individual, departmental, or regional prominence, but rather in a spirit of cooperation, respect, and encouragement.

Fairness: We believe that all employees should be treated equitably and with an open mind, free from favoritism or bias. To sustain an environment where people desire to work and excel, we will recognize performance and address nonperformance.

Like a firm's mission, a statement of its core values is something to be proud of and, as such, should be displayed within your firm as well as to the outside world. Both your firm's mission and—if you choose to articulate them, its values—set the tone for the entire remainder of your strategic plan and your firm's future.

8

YOUR VISION: THE FIRM YOU WANT TO BECOME

"In the long run, men only hit what they aim at. Therefore, you had better aim at something high."

—Henry David Thoreau

It may seem absurdly obvious, but it bears stating: you are much more likely to reach your destination if you know where you are going. While your firm's mission is a statement of its purpose, your firm's vision describes the firm as you *want it to be* at some point in the future—its destination. Before you can move your firm in a defined and positive direction—a direction that you and your firm leaders control—you have to be able to envision and describe the destination to which you are headed.

A compelling vision statement is a powerful and—when stated well—effective tool for a leader to employ. Consider these three historical examples, each of which evoke a strong vision:

- Moses led the Israelites through the desert for 40 years with his ability to instill a vision in their minds of the Promised Land.
- John F. Kennedy mobilized NASA, a relatively new agency, and generated public enthusiasm in the 1960s with his stirring commitment to "put a man on the moon and return him safely to earth by the end of the decade."

- The Reverend Martin Luther King, Jr., inspired an oppressed people with his stirring "I have a dream" speech at the foot of the stairs of the Lincoln Memorial.

Although a design firm's vision will not be as grand as these, it is nevertheless true that when a firm enthusiastically, sincerely, and continually describes its own version of the "promised land" for its employees, it builds morale, motivation, and *esprit de corps.*

THE IMPORTANCE OF VISION

Just as an illustration or rendering of one of your firm's design projects—a building, highway, water treatment plant, or bridge—is a vision of what that project will look like when it is completed, your firm's vision is a conceptual image of what your firm will be like someday. Management consultant and author Stephen Covey advises to "begin with the end in mind," and certainly visualizing the end product makes the process of design much more effective and the probability of success far greater than not knowing what the project will look like when it is complete. No building begins with a roof detail, and no highway starts with a pavement section. The project "vision" gives your client and the end users the confidence that you understand their needs, and helps them maintain interest and enthusiasm throughout the design and construction period. Similarly, your firm's vision is a conceptual representation—in a series of descriptive statements—of the future firm that you will design and construct through your strategic plan. The vision statement gives your staff a clear idea of where your firm is headed and motivates them to all work together toward that destination.

A strong vision statement that is of real value should paint a vivid picture of what your firm aspires to be at a defined point in the future. For most firms, a five-year vision horizon seems to be most comfortable. A shorter period is not enough time to realize an ambitious vision, while a period longer than five years tends to be too difficult for most design firms to visualize with any real clarity.

ESSENTIAL ELEMENTS OF YOUR FIRM'S VISION

In addition to establishing the destination for your firm and getting people excited about the future, your vision helps set the bar for the performance of the firm and the people who are a part of it. A meaningful and effective vision has the following characteristics:

- **It is ambitious, but not unrealistic.** The vision should stretch your firm and the people in it, but it should not be so difficult to attain that it discourages people, or so "pie in the sky" that it inspires cynicism.

- **It is compelling.** The vision should be meaningful and attention-grabbing in order to get your staff excited about the future. And just like the mission statement, it should be unique to your firm.

- **It is multidimensional and descriptive.** The vision should describe many aspects of your firm so that the staff can understand what the whole firm will be like when the vision is realized. Think of the vision as a jigsaw puzzle, with each statement representing a piece of the puzzle, describing one element of what you would like your firm to become. When all of the puzzle pieces are put together, they form a complete picture of your firm in the future.

- **It is shared with everyone inside the firm.** When we talk with staff members—even managers—of firms during a strategic planning process, a common refrain is, "I wish I knew more about where this firm was going." The vision should answer that question clearly.

- **It is confidential.** In contrast to the mission, the vision—like the entire remainder of the strategic plan—is intended only for those inside the firm, especially because it may describe future internal management functions or financial performance.

- **It is led and supported by the leaders of the firm.** Although individuals from throughout the firm at all levels might participate in providing input into the strategic planning process, developing your firm's vision is not a grassroots, bottom-up exercise. It is a major responsibility of the firm's leaders to establish the direction of the firm. There may be debate about the direction during the strategic planning process, but all participants must achieve consensus based on the best interests of the firm as a whole. And once the direction is established, all of the leaders and managers must unanimously and actively support it. As we describe in Chapter 4, sometimes a manager may disagree with the future direction as defined by the vision of the strategic planning team. If this happens, it is much better to recognize and deal with it openly than to ignore or sidestep the dissension, even if it means that one or more of the principals decide that their best future will be outside the firm. The unappealing alternative is to avoid having any vision at all, with conflict continuing to swirl just below the surface. The energy lost in simmering disagreement is worse in the long term than the cathartic experience of getting differences of opinion out in the open and resolving them, even if the resolution is painful in the short term.

STIMULATING VISIONARY THINKING

A good start when developing the vision for your firm is to answer the following questions:

- What kind of firm do we want to become?
- How soon do we want to become that kind of firm?
- What will we look like from the outside?
- How will we operate on the inside?
- How will we measure our progress?

Many architects, engineers, and planners—even firm leaders—tend to be task-oriented people and, as such, find themselves immersed in their day-to-day responsibilities. To help stimulate your thinking beyond your own role in the firm and beyond what your firm is today, we recommend that the firm leaders and managers participating on the strategic planning team consider the following statements and complete them with their own version of the vision for the firm. One of our engineering firm clients refers to this conceptualization process as "visioneering." The final vision may not include all of these items, or it may include items not shown here. Ultimately, the vision needs to work for your firm, so be sure to add or delete questions that allow you to best describe the future that you see for the firm: in other words, your vision.

In the following worksheet, and in selected examples throughout this book, we use the fictitious firm Design Associates to illustrate certain points.

Five years from now, in (20___), Design Associates will:

- Be seen as a (local/regional/national/international) leader serving the following types of clients and target markets: _____

  _____

- Serve clients through offices or a presence in _____
  (Describe target markets and market position; indicate geographic "footprint" and general location(s) of office(s).)

  _____

- Differentiate ourselves from the competition by _____
  (Describe the attributes of your firm that are important to clients and prospective employees and that are unique—that your competitors can't claim.)

  _____

- Be organized around _____
  (Define organizational structure.)

  _____

- Be managed by _____
  (Describe systems of operational management and governance.)

  _____

- Have a corporate culture that _____
  (Express philosophy about intangibles such as what it is like to work here.)

  _____

- Employ a staff of _____ people and generate $_____ in net operating revenue at a consistent profit margin of ____%

  _____

As one example, Design Associates could have developed one portion of its vision that stated:

- In five years, we will be leaders in our target markets of elderly care, hospitality, and commercial developers.

This element of its vision would then drive its decision making throughout the rest of the strategic planning process and the implementation of the plan. Describing the desired future of your firm in these and other terms lets everyone see where the firm is going and allows you to begin thinking in terms of what your firm will have to do to reach its vision.

A Larger Vision: the BHAG

In 1996, Jim Collins and Jerry Porras published a landmark management book entitled *Built to Last: Successful Habits of Visionary Companies*. In the book, they describe a special kind of vision known as a BHAG (pronounced "BEE-hag"), an acronym for Big Hairy Audacious Goal. Not to be confused with quantifiable strategic planning goals described in Chapter 10 of this book, a BHAG is like a five-year vision, only much more ambitious.

- A BHAG is big! It may be so big, in fact, that it may never actually be fully attained. Nonetheless, it compels people to keep striving for it.
- A BHAG is further in the future, 10 to 30 years out. In that way, it keeps people motivated for a much longer time.
- A BHAG is not a sure thing. Collins and Porras note that accomplishing a BHAG requires extraordinary effort, and maybe even a little luck. By their definition, it may only have a 50 to 70 percent chance of success. But, regardless of the odds, nobody in the firm would question whether it is worthwhile pursuing it.

All three of the historic examples of visions at the beginning of this chapter meet the definition of a BHAG. BHAGs are certainly not for the fainthearted and, by our observation, only one in 20—or maybe only one in 50—design firms is ready for and can take advantage of a BHAG. Is your firm one of them?

9

ISSUES: THE ROADBLOCKS TO SUCCESS

"Drive thy business; let it not drive thee."
—Benjamin Franklin

A vision statement describes where you want your firm to be at a specific point in the future, typically five years from the present. If there were no challenges facing your firm, you would be able to move toward that vision unimpeded. In the real world, however, every firm has obstacles, conflicts, and challenges that it faces—its "issues." Your issues will hamper your firm's ability to move from where it is now to where you would like it to be. Issues are the roadblocks that stand in the way of a firm achieving its vision.

Every design firm faces internal issues (those that the firm can manage and control) and external issues (forces that are out of the firm's influence and that, at best, the firm can only anticipate, prepare for, and respond to). Although your firm has the ability to address most of the internal issues it faces, confronting them can be more sensitive and difficult than dealing with external issues.

Issues arise in all aspects of a firm's internal and external universe; the following are common issues that many firms address in their strategic plans:

Internal Issues

- Leadership succession
- Ownership structure and transition

- Organizational structure
- Financial performance
- Financial management
- Staff recruiting and retention
- Staff training, development, and career paths
- Project delivery and project management
- Management conflicts

External Issues

- Economic, demographic, or regulatory changes
- Political climate
- Market shifts, such as emergence, growth, or consolidation
- Technological changes
- Competitor actions

Most internal issues in design firms tend to revolve around people—especially senior people—in the firm and what they do or don't do.

IDENTIFYING ISSUES

A good way to begin identifying your firm's issues is to benchmark your quantifiable attributes and practices against those of similar firms in the industry. Private research firms and professional associations such as the American Council of Engineering Companies (ACEC) and the American Institute of Architects (AIA) publish benchmarking data that allow you to compare statistical survey findings related to financial performance, employee and executive compensation, benefits, turnover, financial management, firm leadership and ownership, marketing, organizational structure, project management, and just about every other meaningful area of concern to firm leaders that can be measured. The better the match—such as size, location, and major services—between the profile of the firms in the statistical sample and your firm, the more accurate and applicable the data will be. Benchmarking a firm's performance against industry medians and upper/lower quartiles points

out a firm's relative strengths as well as the areas where it has potential to improve. Identifying quantifiable areas for improvement flows very nicely into setting goals for your firm, the next stage of strategic planning, which is described in Chapter 10.

SWOT Analysis: Gaining Insight Into Market Issues

A convenient tool that you can use to analyze the impact of certain market-related issues on your firm—and also potential opportunities—is a SWOT analysis (**s**trengths, **w**eaknesses, **o**pportunities, and **t**hreats).

Use a SWOT analysis to evaluate the relative attractiveness of prospective or current services, locations, or markets to your firm. Then, using the organized findings, develop strategies and action items that capitalize on your firm's strengths, take advantage of opportunities, address weaknesses, and position your firm to mitigate threats.

SWOT Analysis Template

Internal Factors

	Strengths	Weaknesses	
Positives	1. 2. 3. 4. 5.	1. 2. 3. 4. 5.	**Negatives**
	Opportunities	Threats	
	1. 2. 3. 4. 5.	1. 2. 3. 4. 5.	

External Factors

Another good way to identify issues that apply to your firm is to ask people inside the firm (both management and staff) and outside the firm (clients, consultants, project partners, and contractors) for their opinions. These groups know what it is like to work for and with your firm, and by surveying them and compiling and comparing their answers, you can identify patterns, trends, and ultimately the overriding themes that emerge from these discussions, which should point directly to your firm's issues. Chapter 3 explains how to conduct broad-based employee surveys (if your firm is large enough) and client perception surveys (a good idea for *every* firm).

If you want to conduct this type of client research through personal interviews, be aware that people tend to speak more freely and raise issues more readily with an objective third party than with someone inside the firm. An independent third party will be likely to elicit much more information and will have the skill and perspective to synthesize it into an objective, coherent, and useful form upon which you can base sound decisions. A good consultant also recognizes a professional duty to represent the information gathered accurately, and, if necessary, to deliver a difficult message without fear of retribution—something very challenging for an insider, who may have to live with the personal consequences of delivering negative findings. It could be that through a consultant's efforts, an issue emerges that no one inside the firm would have had the courage to tell the firm leader, but which the leader absolutely needs to know.

Although facing up to issues may on occasion ferret out some very difficult and sensitive matters related to the performance of the firm leader or another senior manager, we find that the people concerned are rarely completely surprised by the issues uncovered. This lack of surprise may indicate that the leader's awareness of a problem and failure to address it is in itself an issue.

Although many internal issues tend to revolve around people, not all issues are personal. Issues related to areas such as organizational structure, marketing, or employee recruiting and retention are more systemic in nature and typically need to be addressed on a firm-wide basis rather than in relation to the performance of a single individual. We discuss these and other issues that firms often encounter in more detail in Part Three of this book, Chapters 13 through 20.

SYNTHESIZING FINDINGS AND PRIORITIZING ISSUES

We all tend to hear what we want to hear. That's why strategic planning requires some "tough love." Once you have gathered as much information as you can and identified the issues facing your firm, it is very important to compile the findings in a written report to minimize ambiguity and any potential for misinterpretation. The value of the report is in the irrefutable evidence—from management and staff interviews or questionnaires, client surveys, employee surveys, and other sources—that supports its no-nonsense, no-holds-barred analysis. Everyone responsible for developing the strategic plan—the participants in the retreat—should receive a copy of this report, so that everyone has access to the same information at the same level of detail.

After you identify your firm's issues, prioritize them in terms of which have the greatest long-term importance versus which most urgently need to be addressed in your strategic plan. The concepts of "important" and "urgent" are often confused and mistakenly used synonymously. In fact, they are not the same thing, but they are also not mutually exclusive. The word *important* in this context refers to issues that are not immediately time-sensitive but instead are related to long-term investment and success, while *urgent* refers to a short-term necessity. An issue can be important or it can be urgent, or in some cases both.

ADDRESSING ISSUES AND FULFILLING YOUR VISION

The remainder of the strategic planning process can be described as finding answers to two questions, one with a positive connotation and one somewhat negative. First, you must look at your firm's vision, note the gaps that exist between where your firm is today and where you want it to be, and ask, "What do we have to do to make this vision happen?" The sometimes more challenging corollary to that question requires you to take a hard look at the issues you have identified and ask, "How will we solve these problems so that they do not obstruct our path toward our vision?"

Issues That Call for "Tough Love"

The following are some examples of issues faced by real design firms and illustrate the need for "tough love" when addressing issues. We quote the text from management reports that we prepared as the firms' strategic planning consultants:

- **Leadership succession.** "The partners need to identify a successor for Dennis by the end of the year and begin to develop that person into the next managing partner. None of the announced candidates or heirs apparent (Steve, Jean, or Art) have earned widespread staff support and respect. If the partners cannot agree on a successor, their only viable alternatives are to either sell the firm or acquire another firm to provide for future firm leadership."

- **Firm organization.** "The firm's organization structure is unclear and mixes its focus among markets and disciplines. The matrix structure diffuses responsibilities, making it difficult to determine accountability for performance."

- **Growth and staff motivation.** "The firm has not experienced any growth in a number of years. As a result of continued promotions, the staff has become top-heavy and stagnant among its most senior levels. Many of the younger staff members believe that some of the more senior employees aren't really earning their high salaries, and that their own compensation growth potential is constrained because of it. The staff as a whole lacks energy and motivation. There is not a strong work ethic. Nobody is excited to be here."

- **Profitability and financial performance.** "The firm's financial indicators are far below industry medians for similar firms. The net multiplier of 2.84 is much lower than the industry median of 3.31. The utilization rate has plummeted since the end of the fiscal year to 49.8 percent, well below the industry median of 61.9 percent. Profit on net revenue of 2.14 percent is significantly below the industry median of 14.7 percent."

- **Management conflicts.** "There is a feeling that some of the partners are not carrying their weight. The partners avoid confronting one another, which results in the problem of conspicuous non-performance by partners often not being addressed at all."

THE COST OF AVOIDING ISSUES

Although facing difficult issues squarely is not always easy or comfortable, firms risk paying a significant price for not addressing them. Say, for example, that one of a firm's branch offices is performing poorly. Within the context of the firm's vision for its geographic presence, the strategic planning team must determine the causes of the office's poor performance and develop a plan either to return that office to profitability or close it. In this case, failure to address the issue of the branch office's performance could put the remainder of the firm at risk, and at the very least puts high-performing employees, offices, or business units in the position of supporting the unprofitable operation.

Addressing your firm's key issues requires forward thinking: laying out goals to which you can aspire and toward which you can measure your firm's progress, and brainstorming ideas that will push your firm in the right direction.

Why Issues Are So Hard to Face

Most internal issues have to do with the behavior and performance of employees. Issues often arise because their root causes and their solutions are not clearly within any one person's area of responsibility, or because the person responsible chooses to ignore them—or may even be responsible for creating them.

There's a strong human tendency—especially prevalent among leaders of design firms who want to be well-liked—to avoid unpleasantness. But issues, by definition, are unpleasant. The saying "Sugar coating works best on breakfast cereals" certainly applies to design firms when facing their issues. Issues in the workplace tend to persist because most people are more inclined to avoid or work around a people-related problem rather than tackle it directly, especially if it means confronting a colleague or a superior. Even at senior management levels, the greatest challenge that most firms face in strategic planning is identifying and addressing their issues squarely and constructively. Left unaddressed, however, issues can severely affect the performance of—or even threaten the survival of—a firm because they can stand forever as a barrier between the firm and its vision.

Whether they are willing to face issues directly or not, a firm's leaders very likely have, or arguably should have, a strong sense of the issues that are holding the firm back or could potentially hold it back. Unfortunately, that sense does not make it any easier to discuss the issues openly and act upon them.

10

GOALS: MEASURING PROGRESS TOWARD YOUR VISION

"Not everything that can be counted counts, and not everything that counts can be counted."

—Albert Einstein

By defining your firm's mission statement and vision, and then identifying the issues that stand in the way of fulfilling and achieving them, you lay the foundation for defining the future steps in your strategic plan. Now is the time to establish the goals toward which you will assess your firm's progress as it moves toward its vision.

For the purposes of strategic planning, "goals" are quantitative targets at which a firm can aim and toward which it can gauge its improvement. Goals, as they are accomplished, are stepping stones on the path to a firm's vision. They must, by definition, be measurable. Although financial goals immediately come to mind as easy to measure, not all goals lend themselves as readily to quantifying. Nevertheless, if you can attach meaningful numbers to a goal, you will be better able to gauge your progress toward achieving it. As the saying goes, "What gets measured gets done."

The key to defining goals that will propel the firm toward its vision is to set the bar high, but not so high that the goals can never be reached. Goals should be challenging but attainable. In that vein, firms should set goals over a period of one to three years, because numbers attached to anything beyond three years will tend to be increasingly vague and consequently less meaningful.

Goals represent one of the midterm components—like the design development phase in a building design process—that bridges between the long-term, big-picture elements such as mission and vision, and the short-term action plans.

FINANCIAL GOALS

Financial performance is the most obvious area for which a firm can set measurable goals. Anything with a dollar sign in front of it or a percentage sign after it lends itself easily to serving as a target.

A good metaphor for financial performance goals in a design firm is the dashboard of your car: as you are driving down the road, you need to be aware of only the critical functions of your car, for which you need a select number of gauges that you can see and understand separately and collectively, at a glance. Similarly, firm managers—most of whom have technical, not financial, backgrounds—want to be able to read financial reports, quickly and easily interpret the numbers, and make sound business decisions based on their understanding. Adding more and more financial details to reports is like adding more and more gauges and lights to a dashboard; cumulatively, they tend to distract the reader from the most vital statistics in the reports and reduce their apparent importance. Each item of financial data, each individual financial report, or each extra line in a report dilutes the impact of the most critical indicators. Many firm managers have had the experience of receiving inch-thick monthly financial reports, which accumulate unread on their credenzas. You should try to condense your firm's financial goals into the most important or highly leveraged half-dozen upon which the firm's leaders and managers can act.

Key Operational Measurements

There is no shortage of design industry surveys, which are chock-full of every financial performance metric imaginable. But very few design firm leaders have the inclination or patience—much less the time—to analyze their firm's performance to the nth degree. Using the dashboard comparison—and some firms even produce what they call a "dashboard report"—here are some of the most essential "gauges" by which you can evaluate your firm's performance:

- **Sales**. A "sale" is a signed contract, and nothing less. Sales contribute to backlog (until the work is completed and invoiced) and provide the fuel for your firm's always-hungry engine. Bear in mind that not all sales become real work (even some projects with signed contracts don't fully—or even ever—materialize; or the project may be delayed).

- **Net revenue**. This is all the money your firm takes in (as gross revenue) minus what goes "out the back door" for paying project subconsultants and direct reimbursable expenses. In other words, it's the money you get to use for running your business, keeping, distributing, and investing what's left over as profit.

- **Profit margin**. When you subtract your employees' total payroll costs, both direct labor chargeable to projects and nonchargeable indirect labor, along with other *indirect* expenses of your business, what's left over is profit. Divide your firm's profit dollars—before you distribute discretionary bonuses or pay taxes—by its net revenue dollars to get your firm's profit margin stated as a percentage.

- **Utilization (or chargeability)**. Divide your firm's direct chargeable project labor cost by its total labor cost, and the resulting percentage is your firm's utilization. Because most design firms essentially sell their time, higher utilization usually translates into higher billings and greater potential profitability.

- **Net multiplier**. Divide your firm's net revenue by its *direct* (chargeable to projects) labor cost, and the result is your firm's net multiplier. This measures the value of your firm's services to its clients (the net revenue numerator of the ratio) and the efficiency with which your firm executes its work (the direct labor cost denominator of the ratio), so the higher the net multiplier, the better.

- **Revenue factor**. Divide your firm's net revenue by its *total* labor cost to calculate its revenue factor. Like the net multiplier, it measures the value of your firm's services in the numerator and the efficiency with which your *entire* firm—technical project staff and nonbillable staff—operates, because its denominator includes *all* of your firm's labor costs, not just direct project labor. As with the net multiplier, a higher revenue factor is better.

- **Backlog.** Your firm's backlog is the total amount of work under contract that has not yet been completed or invoiced. Divided by your firm's total monthly payroll, backlog can be measured in months of work. Because new sales do not always materialize into "real" projects, some firms conservatively calculate their backlog to include projects for which they have gone through one monthly billing cycle.

NONFINANCIAL GOALS

There are many areas other than financial performance around which firms should set goals. The following are examples of some nonfinancial goals that can be set and measured:

- **Growth in staff size**. *By [a date one to three years from now] we will increase our staff size by 25 percent to 65 employees.*
- **Employee recruiting and retention**, measured by the number of job candidates attracted and résumés received, the number hired, and the firm's voluntary and involuntary turnover rates. *We will achieve an annual turnover rate of no higher than 10 percent of our total staff.*
- **Geographic expansion**, measured by the number and/or locations of offices or by the market share the firm occupies in its target geographical markets. *We will open two new offices, via acquisition or with our own resources, one in Missouri and one in Illinois.*
- **Staff development**, measured by the number and types of accreditations earned by employees: professional licensure, sustainable design accreditation, or any other specialized accreditation that is appropriate to the firm, its professional disciplines, or its target markets. *Twenty-five percent of our professional staff will apply for accreditation as Certified Floodplain Managers.* Or: *All of our project architects with three or more years of experience will have completed our project management training program by the end of their fourth year of service.*
- **Marketing success**, measured by the "hit rate" from proposal to short list, and from short list to win. *We will apply our go/no-go protocols to every marketing pursuit and convert 75 percent of our proposals to short-listing.* Or: *We will develop five unsolicited proposals or white papers per year to submit to state and federal clients.*

- **Ownership transition**, measured by the percentage of ownership that is distributed among a defined number of individuals, or by the percentage of ownership that is transferred over time. *Charlie will reduce his ownership from 60 percent to 45 percent over the next three years and no future owners will own more than 20 percent of the firm.*

Although all of these example goals can be expressed in numbers, each is related in some way to reaching a firm's vision or would enable the firm to address one of its issues.

"Shrinking" Goals

While it may at first seem counterintuitive, especially to firms that aspire to growth in some way, not all goals have to involve *increasing* numbers. For example, a 100-person A/E firm had 12 principals, all of whom had varying shares of ownership in the firm, and by virtue of their ownership status, each was a member of the firm's board of directors. During the strategic planning process, it became clear that a board of this size and composition stymied decision making. There were too many participants, and not all of them had the proper skills to lead the firm (though each of them was valuable to the firm in other ways). Making every owner an equal member of the firm's board also set an unsustainable precedent for broadening ownership. Therefore, the firm decided to *shrink* its board from 12 to 7 and ultimately to 5 members through a series of elections by the then-12 shareholders. Similarly, goals related to growth in target markets may direct a firm to place more emphasis on certain markets and less emphasis on others.

MAXIMIZING THE VALUE OF GOALS

Firms should update their goals annually so that the three-year, midrange goal "horizon" between today and the firm's five-year vision is continually moved one year further into the future. Firms should also publicize their goals widely among the staff so that everyone understands them and has a personal stake in helping the firm reach them.

Sharing the firm's goals with all staff members usually means sharing financial information, an idea that firm leaders sometimes resist. The most common rationale we hear for their reluctance to share financial goals with all employees is that "the staff will not understand them." Although it is true that some employees may not take notice, show interest, or even have the aptitude to understand financial metrics, many more will. Withholding this information from employees who are interested in and have the ability to understand financial information on account of those who do not is a form of "management by the lowest common denominator" and has the effect of dumbing down the firm. It deprives the entire staff of the knowledge and insight into the management of the firm that many employees could otherwise gain and impedes the development of the future leaders for the firm. If a firm leader wants everyone to be walking the same path toward achieving the firm's goals and vision, everyone needs to know where the firm is headed and how it is progressing.

For example, if a firm sets a goal to reach a certain net revenue level within three years, and that goal is further broken down annually by business unit (studio, department, office, etc.), then each employee in a given business unit can more readily see how his or her work contributes to the bigger picture by being allowed to see and understand the firm's financial reports. Conversely, if those same employees are kept in the dark about their firm's or business unit's goals, they are likely to be less motivated to help the firm reach those goals. There's another important benefit to sharing financial information with employees: understanding the "business side" of the firm will someday make them better managers of your company.

Celebrating Your Achievements

By all means, celebrate when you reach your goals. Many design firms that set goals do not celebrate their accomplishments enough. It is important to take a step back from tasks and projects often enough to view the big picture. Publicly acknowledging the accomplishment of a goal—or even making progress toward it—has a very positive impact on morale. Removing certain members from a firm's board of directors might not be an action that a firm would want to celebrate, but the successful reorganization of the board into a more effective decision-making body certainly would be.

Our fictional firm, Design Associates, had the following statement in their vision, "In five years, we will be leaders in our target markets of elderly care, hospitality, and commercial developers." As an outcome of that five-year vision, the firm set shorter three-year goals around the revenue in each of their three target markets:

Revenue Goals by Market	2008	2009	2010
Elderly Care	36%	40%	50%
Hospitality	27%	30%	30%
Commercial Developers	30%	25%	20%
Other	7%	5%	0%
Total %	100%	100%	100%
Total Net Revenue ($ mil.)	$6.9	$7.5	$8.2

Meeting goals will not happen by itself. Your firm needs a push in the right direction, and this push will come from strategies that you will develop as the next phase of the strategic planning process.

11

STRATEGIES: PUSHING YOUR FIRM IN THE RIGHT DIRECTION

"If you want something you haven't had before, you have to do something you haven't done before."

—Bill Wilson

The word *strategies* can be misleading. After all, this entire book is about strategic planning. Yet strategies themselves are only one component—albeit an integral one—in a complete strategic plan. The term for which strategic planning is named, strategies are the great ideas that push a firm in the right direction, describing just what it should do to achieve its goals, address its issues, and ultimately achieve its vision.

Strategies should be—and often are—relatively easy for design firms to develop because the people who manage design firms tend to be creative problem solvers, brimming with ideas. Therefore, the best way to generate strategies is by capitalizing on this collective creativity through group brainstorming. When faced with an issue or an unfulfilled vision, ask, "What are we going to do about this?" (Or, adding a necessary dose of reality, "What are we *willing* to do about this?") As famed research scientist Linus Pauling said, "The best way to have a good idea is to have a lot of ideas."

In the strategic planning process, there are often two opposing forces that must be reconciled in the strategic plan itself: positive forces that take advantage of opportunities and propel a firm toward its vision, and negative forces that would hold the firm back if they were left unattended. Similarly, a strategic plan can contain two types of strategies, depending on whether a firm sees the glass as half full—an opportunity to capture—or half empty—a problem to solve.

In the half-full approach, the firm asks, "What can or should we do to fulfill our growth vision?" A strategy might be "*increase our investment in marketing to 6 percent of net revenue.*" In the half-empty scenario, the firm asks, "What can or should we do to address our key issue regarding poor cash flow?" A strategy might be "*increase the frequency of our invoicing.*"

If a plan includes only half-full strategies, it risks being overly optimistic or perhaps even naïve. If a plan includes only half-empty strategies, the people in the firm can come to see the plan as a negative process, one that beats up the firm and demoralizes them. The best strategic plans include both types of strategies, each complementing the others and all together creating an ambitious and still realistic plan.

DEVELOPING STRATEGIES

Asking broad, open-ended but provocative questions is a good technique to brainstorm new strategies:

- What could we be doing that we've never done before?
- What should we be doing more of that's working well for us now?
- What should we cut back on that's not working for us now?
- What should we stop doing that's been shown to be counterproductive?

In previous chapters, we used the fictional firm Design Associates and its strategic plan as an example. Design Associates' vision called for the firm to "…be leaders in our target markets of elderly housing, hospitality, and commercial developers." The firm subsequently identified issues—including a current lack of elderly housing market penetration and capabilities—that constrain its ability to reach its vision. After setting three-year annual revenue goals for the elderly housing market, the firm developed the following strategies to allow it to reach its goals:

- Increase investment in marketing to elderly care clients to 5 percent of net revenue
- Recruit senior-level staff with elderly housing design experience
- Increase production capabilities in anticipation of an increase in elderly housing workload
- Investigate potential acquisitions of firms in the elderly housing market
- Develop specialized services that the firm could provide to elderly housing clients

You can see that each of these five strategies represents a good idea that should move Design Associates closer to its goal and vision. Conversely, you can also see that if the firm did not refine these ideas to a greater level of specificity and accountability, it would be far less likely to achieve its goals and vision.

When Is a Strategy Not a Strategy?

When you are setting goals for your firm and strategies and action plans to achieve them, you might be confused as to how to differentiate among the three terms. Goals are measurable, strategies are general ideas, and action plans are specific tasks. Let's review examples of each.

Goal	Strategy	Actions
Employ a total staff of 45 people	Improve recruiting and retention of engineering staff	Chris to participate in two top-ten engineering school job fairs next fall Joan to compile database of employees who have turned down a job offer from us or left our firm in the past two years by June 30.
Increase gross revenue by 25%	Enter one new market	Frank to research transit, urban renewal, and historic preservation markets and make recommendation to the board of directors by April 30.
Expand throughout the Midwest with a presence in one new state	Open an office in St. Louis	Sue to work with executive search firm and make strategic hire to lead St. Louis office by September 1.

See Chapter 10 for more information about goals; Chapter 12 describes action plans.

STRATEGIES AS INVESTMENTS IN CHANGE

As integral components of the overall strategic planning process, strategies involve change. If a firm wants only to maintain the status quo, there is no need for strategies and, in fact, no need for a strategic plan at all. Because strategies involve change, they almost always require an investment of some sort: money, time, attention, energy, or, most often, a combination of these. Unfortunately, though design firms can be adept at brainstorming new strategies, when it comes time to put them into action, many firms resist the investments—especially the monetary and time investments—necessary to realize their strategies.

Some of us who grew up in the era when cars had carburetors (versus computerized fuel injection systems) recall that it was possible to manually adjust the fuel/air mixture to the engine. Even without a tachometer or fancy tools, with the engine idling we could slowly turn the needle valve on the carburetor until the idle speed peaked and the fuel/air mixture was optimized. The optimal mixture would not be the leanest with the least amout of fuel going into the carburetor, because though that may appear to save fuel in the short term, it would not allow the engine to run at peak efficiency. It is the same with investing in strategies for a design firm. Just as "starving the engine" did not achieve the best automotive performance, shortchanging the investments necessary to make strategies happen may appear to save money, time, and energy in the short run, but it will hurt the firm over the long term. To make strategies work—and they have to work for strategic planning to succeed—firms must allocate the right mix of resources.

ISAAC NEWTON AND STRATEGIES

In his first and second laws of motion, Sir Isaac Newton concisely and presciently described how design firms deal with strategies. Newton's first law states, "An object at rest will remain at rest unless acted upon by an external and unbalanced force. An object in motion will remain in motion unless acted upon by an external and unbalanced force." Translating this theory from "objects" to design firms, we have noticed that managers tend to resist new strategies for a number of reasons, most of which relate to Newton's first law. We often near the following reasons—actually excuses—not to change:

"This is the way we've always done things." (an "object" in motion)
"Our firm has never done that." (an "object" at rest)

This brings us to Newton's second law, which states, "The rate of change of the momentum of a body is directly proportional to the net force acting on it." If strategic planning is intended to change a firm (for the better, of course), then a "net force"—a strategy—is required to act on the firm. The more deeply entrenched the firm is in doing things the same way, the more force will be required to change the firm's direction.

CHANGING STRATEGIES

Nothing stays the same. Markets change, clients change, people change, and technologies change, so firms have to change. The good ideas that form a firm's strategies are not—and cannot become—stuck in time. Firms have to adapt to current and future realities to continue to push the firm in the right direction. As firms update their strategic plans, usually on an annual basis, they should reexamine their strategies to make sure that they still make sense, that they are still grounded in reality, and that they are still sufficient to "move" the firm. Rarely if ever will updated strategies require wholesale change; more often they will need a bit of tweaking and fine tuning, with a few seeming to have become less relevant and a bit of renewed brainstorming being required to generate new and more potent ideas.

Strategies push your firm in a direction, but if the strategic plan ended with strategies, the planning group would leave the retreat not knowing just what to do next and who was to do it. That information forms the next and most detailed component of a strategic plan: action plans.

12

ACTION PLANS: THE MEANS TO REACH YOUR GOALS

"Plans are only good intentions unless they immediately degenerate into hard work."

—Peter Drucker

Action plans are the most detailed component of strategic planning. To bring the metaphor home to the design process, action plans are the "construction documents" of the plan. Imagine a contractor trying to build one of your firm's projects without adequate construction documents, and you'll understand how inadequate and incomplete a strategic plan would be without action plans.

Action plans describe who is going to do what and by when. They consist of a series of specific action items to be completed in the short term—usually the 12 months following the planning retreat—to implement the strategies, achieve the goals, and ultimately reach the vision that the firm has adopted. In this way, they help you to move your firm one step at a time toward your vision.

Action plans have three necessary parts:

1. A **specific action item** with a beginning and an end so that everyone understands what is to be done

2. The **name of the person** who is responsible for accomplishing the action
3. The **date** by which the action is to be completed

Let's discuss these three components of action plans one at a time.

DEFINING ACTION PLANS

Action plans are derived from strategies by asking, "How can we accomplish this strategy?" In this way, one strategy may require multiple action items. Action items are not open-ended. For example, "address scope changes from clients more promptly and assertively" is not an action, but rather, a strategy or "good idea" (see Chapter 11). Likewise, "make our weekly project scheduling meetings more efficient" is also a strategy. Actions have a completion milestone or definitive event that differentiates them from ideas (strategies), and allows the person responsible for that action—and everyone else—to know when the action item is complete and to be aware of progress toward completion.

An action can involve the following:

- an **event or activity** such as a strategic hire or the opening of an office
- an **investment or purchase** such as a new software program or piece of equipment
- a **deliverable** such as a market research report with recommendations, a marketing plan, or a database

Because action plans are so specific, there should be no ambiguity as to when an action has been completed. Also because of their specificity and the deadlines associated with them, there is real individual accountability associated with taking on action items.

RESPONSIBILITY FOR ACTION ITEMS

Every action item needs a person responsible for accomplishing it. That individual should be thought of as the action's "project manager."

Just as with a design project in a firm, the project manager rarely takes on the entire project alone. Instead, he or she relies on a team to get the work done. In carrying out action items, the individual responsible for completing each action can—and often should—rely on others for assistance. In fact, one of the best ways to broaden a firm's participation in its strategic plan beyond those select people in the retreat is to ask for volunteers (or occasionally, to draft people) to work on the actions. This relieves some of the burden from the person responsible for the action item and also allows others in the firm to have a role in the plan and develop a deeper feeling of buy-in and gratification that the plan is theirs, too.

There are three key items to keep in mind when determining who is to be responsible for an action item:

1. **Participation.** The person must be at the planning retreat. Because the person participated in the discussion that led up to the action, he or she has a much greater understanding of the thinking behind it, its intent, and what completing it will involve. Plus, it is unfair for the managers of a firm to sit in a planning retreat and delegate responsibility for an action item to someone who isn't there. In fact, a good ground rule for "assigning" an action item is that if nobody in the planning group will take on responsibility for it, it shouldn't be part of the plan. Without a "project manager" to "own" it, the action item is destined to fail because the firm's management apparently lacks the will to make it happen.

2. **Accountability.** Each action item should have only one person—its "project manager"—responsible for its completion. This contributes to much-needed accountability for getting things done and prevents the "I thought he was doing that" scenario, in which the leadership for an action item is ambiguous. There may be occasions when parallel actions are consolidated, with multiple people working together on them, such as several market leaders each developing their own marketing plans simultaneously. But even then, each of the market leaders is clearly responsible for his or her deliverable. Again, there is great specificity and no "wiggle room."

3. **Workload.** No one should be assigned more than he or she can handle. In the zeal of a planning retreat, it is easy for a few enthusiastic or broad-shouldered people to assume responsibility for too

many action items. Avoid this, not just because putting too much on anyone's plate lessens the likelihood of accomplishing everything, but also because it is an educational and career-building experience to take on actions from a strategic plan. Although assigning action items to certain individuals will be a "no-brainer" (such as directing financial action items to the CFO), the planning group should make sure that the remaining actions are spread out among the group to the extent possible.

ESTABLISHING COMPLETION DATES

Every project for which a design firm is hired has deadlines. Clients insist on deadlines because a project schedule allows them to monitor progress, plan other activities that are dependent on the deadlines being met, and hold the design firm accountable for meeting its deadlines.

So it is with action plans. Every action item is bounded by a deadline. Describing the action and assigning (or volunteering) the person responsible accomplishes nothing if there is no fixed date by which it is to be completed.

THE REALITY CHECK

The test of an effective strategic plan is not how many action items the planning group can generate. Our experience has shown that having too many action items only serves to dilute the time and attention that the firm can devote to the more important action items. Therefore, the planning group should rigorously prioritize the action items, selecting only those that are most important for the firm to accomplish.

At the end of every planning retreat, the group should take a big step back, look at their work on the flipchart pages papering the walls, and ask themselves, "How many action items can our managers reasonably take on in addition to our full-time jobs?" If strategic planning is to be an ongoing activity punctuated by annual retreats, those action items that didn't make the "hit list" this year will have their chance next year, after this year's action items have been completed.

A Method for Monitoring Action Items

Your strategic plan will include specific action items assigned to various "project managers" in your firm. The firm leader should monitor progress on action items and hold the "project managers" accountable for their completion. The following method works well:

- Create a master list of *all* action items in chronological order.
- Create individual to-do lists (for each person in the planning retreat who has an action item assignment).
- Use the master list as the agenda for regular meetings of the strategic planning group.
- If an action item "project manager" misses one deadline, reschedule it. If he or she misses two deadlines, immediately reassign the action item to someone else. Getting the action item accomplished for the good of the firm is more important than who accomplishes it.

MEANINGFUL ACTION PLANS

Our fictional firm, Design Associates, has developed strategies to increase its presence in the elderly housing market. To put those strategies into motion requires specific action items such as the following:

Elderly Housing Actions	Who	When
Research and develop a list of top elderly housing clients within 50-mile radius of the office	Adam	Jun 1, 2007
Develop a marketing plan for the elderly housing market	Julia	Sept 1, 2007
Develop a list of criteria for hiring a senior elderly housing designer including job description, skills, and salary range	Julia	Oct 1, 2007
Identify a list of nontraditional design services that could be provided to elderly housing developers and operators	Sam	Jan 1, 2008

There are several items to note about the example above. First, the simple table format allows easy review or sorting by action item, "project manager," and deadline. Second, displaying the action items in chronological order by due date allows the list to serve as a time-defined agenda for progress meetings. Finally, each action item has just one person responsible for it. Julia has two action item assignments, which is appropriate because she is the market leader for elderly housing and would logically be responsible for developing the marketing plan and hiring a senior designer.

Part Three

STRATEGIC PLANNING ISSUES

13

BUSINESS GROWTH

For many firms engaged in strategic planning, growth tends to be a hot-button issue for one of two reasons: either the firm is not growing (or not growing enough) or the firm is experiencing the wrong kind of growth, often described by overworked managers as "uncontrolled" growth. A commitment to growth is a vital component of any strategic plan, not because every firm should grow for growth's sake, but because firms simply must grow in order to remain successful in competitive markets, and *all* design firms exist in competitive markets.

Growth may be defined in terms of sheer size—as measured by revenue or number of employees—or greater profits, or, alternatively and more intangibly, in terms of higher quality work, more interesting projects, or more desirable clients. Although these professional areas of growth are worthwhile objectives, they are difficult to achieve without intentionally or consequently increasing the size of the firm and establishing a greater presence in its markets. For this reason, we define growth in the most fundamental and universally understood way: by firm size in terms of number of employees and revenue.

WHY FIRMS RESIST GROWTH

When design firm leaders are comfortable with the current state of their firms and their ability to manage them, they may be ambivalent about—or even resistant to—the idea of growth. They are satisfied with the volume of work in the firm, the types and number of projects, the composition of their management team, the number of employees and offices, and the firm's current revenue and profitability. If the firm has been performing well financially, they may be very content with their own total compensation. They may perceive growth as a risky or costly venture that will disrupt existing business operations, working relationships, and profitability. But often, the reasons behind the arguments against growth are simpler and much more personal: fear and insecurity. The firm's leader or senior managers may fear that the business will grow beyond their ability to control or manage it, or that a growing enterprise could afford opportunities for others in the firm to surpass them. Growth may require more work, possibly combined with a greater level of accountability on the part of managers or even the firm leader.

Some of these fears could be well-founded. The leader of a 20-person firm may be able to fulfill his or her leadership responsibilities while still being actively involved in projects and maintaining a high level of personal contact with every client. The leader of a 50-, 100-, or 200-person firm has to maintain more external focus on the overall direction of the firm and oversee a more complex management structure, and therefore exercise leadership, management, and interpersonal skills on a different scale than the leader of a significantly smaller firm.

Growth affects managers at all levels of the organization. Whatever a person's role in a firm, the prospect of having to change the way the firm operates may push people beyond their comfort zones. Therefore, firm leaders and managers must engage in frank discussions about how roles and responsibilities—theirs as well as others'—need to evolve as the company grows, and how they can support one another to fulfill them.

The larger the firm, the greater the distance between the firm leader and the day-to-day activities of projects. For many design professionals whose passion and skills lay in the art and science of their profession, the transition into management may have been difficult, even painful. And at some point, a growing firm may even require specialized leadership and management skill for which a technical design education alone may simply be inadequate. When facing the question of growth, courageous leaders ask themselves, "Is growth really more of a risk for the firm or for me?" Visionary leaders—like mature, responsible parents—do not fear the growth and development of their progeny. They cultivate it because, in the end, it is the most viable—and also the most gratifying—strategy for everyone involved with the firm.

THE COMFORTING ILLUSION OF THE STATUS QUO

Like many other types of businesses, design firms exhibit the traits of living organisms. No living creature is designed to maintain a status quo. At any moment in time, living things are either continuously growing and developing, or withering and decaying.

The situation is very much the same for a business, including design firms. At any moment in time, a business is either on an upward trajectory or a downward slide. There is no validity to the concept of "business equilibrium." It is almost impossible for any business to maintain the status quo for any length of time: the same number of employees; the same clients and types of projects; the same revenue, expenses, and profit; the same cash flow; or the same level of competence and skill. It's inevitable that some employees, clients, and projects will come and go. Business and market conditions will change. Costs will increase. Technology will advance. And while a firm is preoccupied trying to maintain balance, competitors will be busy trying to woo away its clients and recruit its employees. Unless the firm is focused on growth, it will fall behind because the rest of the world is simply not standing still. The cumulative skill sets, business practices, and marketing strategies of a firm all need to be sharpened constantly and refreshed just to retain the firm's current market share. A commitment to growth is not only key to achieving a greater level of success, it is essential to a business's survival.

GROWTH BENEFITS EMPLOYEES

In any labor market, the need to retain increasingly qualified employees demands that firms accommodate and facilitate the employees' career ambitions. Therefore, employees at any level in a firm need career development and growth opportunities, whether or not they all avail themselves of them. Some employees may be content with their current responsibilities and day-to-day work, but a growing firm affords even those employees the opportunity to apply what they've learned from one project to new (and perhaps more challenging) work. More importantly, firm growth allows ambitious employees—who are more vital to the business's future—to develop, learn, take on more challenging assignments, improve their technical and management skills, and advance in their careers. A growing firm allows *people* to grow, to develop professionally. It's the very rare boutique firm that can provide significant technical, professional, or artistic challenges to its employees without growing, and in those few firms, employees typically come and go with great frequency, often just to "study at the feet of the master."

A long-term vision that includes the growth of the firm sends a clear message to all employees that they each will have the opportunity to advance on an individual level; that as the firm grows, opportunities will become available to work on larger or more interesting projects, or manage people or even offices. A firm that does not articulate growth as a goal sends the opposite message to employees, who quickly and unfortunately come to understand the dynamics of stagnation: a dull repetition of work, and a group of people entrenched above them blocking their career paths. If a 50-person firm has three principals who share the ownership of the firm, each of whom is years away from retirement, and the firm has no vision or plan for growth, everyone in the firm will recognize that their career prospects with the firm are limited. Regardless of how favorable all other aspects of employment may be, this one factor can have a strong negative impact on morale, productivity, and the ability to retain the best staff. Ultimately, the best, most valuable employees will leave.

There is also a relationship between firm size and employee compensation. Economies of scale often allow larger firms to compensate employees more generously. Salaries tend to be higher in larger firms because larger projects with higher fees can translate into greater responsibility for otherwise equivalent positions at all levels. And the higher the position in an organization, the greater the difference in salary is likely to be between smaller and larger firms. Industry surveys show that managers—principals, project managers, branch office managers, CFOs, marketing directors, human resources directors, and IT directors—generally earn higher median base salaries in larger firms. Larger projects with larger fees, larger project teams, and larger business units all tend to translate into higher compensation for the people responsible for managing them.

Of course, firm growth also provides opportunities for professional growth other than through increased management responsibilities. Larger project size and greater project complexity create opportunities for increased technical specialization and career growth along a technical rather than a management track. While many firms wrestle with the "dual track" phenomenon of career advancement and few firms formalize the distinction between the two, smaller firms simply aren't able to provide their employees with the same range of specialized career options as larger firms.

The relationship between firm size and profitability is less clear. Some small, highly specialized firms are highly profitable. Factors other than size have a greater impact on design firm profitability: market focus, geographic location, internal operations, and business management. Because shareholders of smaller firms—which typically have more concentrated ownership—tend to derive a larger share of their income from profit distributions instead of salaries, the reported prebonus profits of smaller firms tend to be higher than for larger firms, making accurate statistical comparisons between firms of different sizes difficult.

GROWTH ENHANCES FIRM VALUE

Growth is as important for the firm as it is for the individuals in the firm. The intangible worth of a firm is in its cumulative intellectual capital. A firm cannot improve—or even maintain—its competitive advantage in the marketplace unless it continually cultivates the intellectual and professional development of everyone in the firm. But intellectual capital is an asset over which a firm and its leaders have no absolute control—it rides the elevators and walks out the door every night, as the saying goes. The larger the firm, the broader the foundation of that capital, and the less dependent the firm is on particular individuals for its success.

A firm that is not growing cannot afford—or justify—investing heavily in marketing, technology, or developing the capabilities of its staff. A growing firm provides its employees with professional development opportunities in the best, most cost-effective and productive way: by providing them with a continuous flow of ever more challenging work.

The Increasing Value of a Growing Firm

Although growth benefits both employees and your firm, there are added benefits of growth that accrue primarily to your firm's shareholders. Your firm's value is a function of its financial performance. But how does the marketplace determine the value of privately held companies, such as most design firms? Potential investors, whether they are *internal* or *external* buyers, will look very closely at their projected return on investment (ROI). Past trends in revenue growth and profitability will figure very prominently in the calculation, because the historical trends are strong indicators of future performance. When all is said and done, a firm's forecast of future earnings is simply the projected profit margin times its revenue base; and if the revenue and/or the profit margin are projected to increase, so will the value of the firm.

For the managers and employees who would most likely buy out the current owners of a professional services firm, growth trends are particularly important. These internal buyers typically do not have large amounts of capital, and ownership transitions tend to occur over an extended period of time. The future owners may rely heavily on projected revenue growth and profits to determine the value of their purchase, and thereby their ROI. In the absence of a positive culture of growth, the investment may seem too risky to them, and they may opt not to buy. Also, as Chapter 17 describes, internal buyers will likely rely on their share of the firm's growing profit to help them pay for their new shares. A shrinking pool of likely buyers has a negative impact on the value of a firm. Whether you are a current or future owner of a firm, a policy of growth can only enhance the future value of your shares.

PLANNING AND MANAGING GROWTH

In and of itself, size will not lead to success, nor will an uncontrolled growth-through-diversification strategy. Whether a firm's ambitions are local, regional, national, or global, its growth strategy should be aimed at capturing a significant—or better yet, a dominant—share of a defined market or markets centered around specific types of clients. The best market opportunities are rooted in a *focused* business strategy, because no firm can be all things to all clients.

At first, this may be hard to accept. After all, architects are educated to believe that they can design anything from a brochure cover to a skyscraper. Engineers are taught to be technical problem solvers, regardless of the context of the problem. Planners are taught to plan parcels, subdivisions, and even entire cities. Although such self-confidence in one's abilities is admirable, it provokes tremendous anxiety in clients who are investing real money in real projects in the real world and who want to mitigate their risks. They do this by seeking out specialists: people who know what they're doing, who have a track record of success on similar projects, who understand how their clients think and what their clients need, and who won't reinvent the wheel—at least not on the client's dime.

As Chapter 14 discusses in the context of organizational structure, firms with a strong market focus are among the most successful because of the specialized knowledge, project leadership, and, hence, value that they deliver to their clients. This translates into premium fees, which, when combined with the knowledge and efficiency that come from focus, often translate into greater profitability. Although design professionals are educated to be generalists, the marketplace places the greatest value on their specialized skills, including their ability to understand their clients' problems and priorities. There are a few exceptions, of course, particularly in architecture. A "signature" architect may get a commission to design a museum, concert hall, or airport even if he or she has never designed one before (always backed up by an experienced production firm), but most firms would be ill-advised to pattern their business strategy after these few exceptions, however high profile they may be.

Many firms wrestle with the issue of whether they should focus on a single market or on a few markets because narrowing one's focus seems counter to a growth strategy. Some managers naively believe that their firm's future success depends on its ability to diversify without limits. However, it would be a mistake to equate managing a design firm with managing an investment portfolio. The intent of diversifying investments is to spread and minimize risk. But as an investor, you expect each individual investment in a diversified portfolio to be managed by a recognized expert whose goal and obligation is to maximize the return on your investment. Similarly, clients of design firms minimize their risk by seeking out experts for each type of project. It is almost impossible for a heavily diversified design firm to cultivate a legitimate reputation for expertise in a large number of markets, maintaining the necessary depth and breadth of expertise. This is particularly true for the many small to midsize (30- to 60-person) firms. A diversified or generalist midsize firm with only a relatively few number of people dedicated to any one market will rarely be perceived as the go-to firm in those markets and will always be fighting an uphill battle against its more market-focused and specialized competitors that can demonstrate greater depth and specialized knowledge.

Consulting a Specialist

A firm will enjoy greater success developing a focused growth strategy if its leadership can think like a client or consumer rather than like a provider of services. If you have a serious medical ailment such as a chest pain, would you entrust your care to a group of doctors who advertise as "one-stop" specialists in cardiothoracic conditions, orthopedics, neurosurgery, and podiatry? Not likely. The more significant your illness or acute your symptoms—in other words, the more that's at stake—the more comfort you take in your physician's specialized knowledge.

UNDERSTANDING MARKETS

Consumers of design services—whether they are homeowners, real estate developers, state transportation departments, federal agencies, schools, or corporations—have similar concerns. Every project presents a significant risk. In some cases, the money at risk belongs to others—investors or taxpayers. There are jobs, reputations, and even careers on the line. The typical client wants knowledgeable specialists for his or her projects: design firms that understand their clients inside and out; that live, sleep, eat, and breathe their clients' issues; and that instinctively look at projects from their clients' points of view.

A market for design services is a group of clients who think alike, share the same concerns and priorities, and have the same goals. A firm can market by communicating with this group of clients in a language they all understand, discussing issues to which they can all relate (Chapter 18 discusses this in more detail).

Design firms often make the mistake of lumping markets together into convenient categories by project type or geographic area, but the people in these categories may have very little in common. The following table illustrates the difference between categories and markets:

Categories *Nothing in Common but a Label*	**Markets** *Common Interests, Concerns, Needs, Priorities*
Education	public K–12 schools private K–12 schools public colleges and universities private colleges and universities
Geographical Area (city, county, region, state)	public recreation facilities individual homeowners neighborhood associations places of worship small business owners
Government	airport authorities county corrections departments county courts state courts federal courts local housing authorities local police, fire departments military state transportation departments
Health Care	acute care/specialty hospitals ambulatory care centers assisted living facilities community hospitals research/teaching hospitals nursing homes medical office building developers
Real Estate Developers	residential developers/large scale homebuilders office/commercial developers retail developers hospitality developers

MARKET FOCUS

A good and frequently cited example of a design firm with a highly specialized market focus is HOK Sport (as the firm is best known), a division of the firm Hellmuth, Obata, and Kassabaum, Inc. Although HOK Sport may not win the commission to design every major sports stadium in the world, the odds are very high that it will be on every short list. In the minds of people who build sports stadiums, HOK Sport is the recognized leader in stadium design. Indeed, it could be argued that the firm invented the market and branded itself by recognizing that there existed a distinct group of people who think alike, share the same concerns and priorities, and have the same goals, and *whose needs were not previously being met adequately by an obvious market specialist.* Today, any team or jurisdiction building a stadium is virtually compelled to seriously consider HOK Sport for the project. That gives the firm a tremendous advantage in the marketplace that few other firms have.

HOK Sport is not alone in its market; it has considerable competition from firms of similar size, experience, and expertise. But how many of these competitors can you readily name? More importantly, how many can prospective clients name? The difference between HOK Sport and some of its competitors is one of laser-like business focus, supported by effective branding (which is much easier and more effective with that level of focus). Market focus is a strategy of narrowing a firm's business with the intent of dominating a market. When you dominate a market, your firm name becomes a brand that is synonymous with the market, especially if the name has marketing appeal. Brands, once they are lodged in the minds of their customers, are extremely difficult for competitors to dislodge. By creating the HOK Sport brand, the firm signaled to professional sports leagues, team owners, and cities its total immersion in and single-minded dedication to the market, which it supports by delivering high-quality, specialized design services. To become the go-to firm in any market for design services, a firm has to have three attributes of market focus: a visible leadership position in the design specialty, a critical mass in the market, and the ability to deliver.

FOCUSED GROWTH

"Market focus" does not mean that design firms should focus on just one market. But a firm with a strong market focus should guard against unfocused, uncontrolled growth, and instead use its secure and profitable position in one market to expand into related markets through *selective* diversification.

On how many markets can a firm maintain a legitimate focus? There is no definitive answer, because every firm has its own limits of capability and capacity. The defining variable is the ability to develop critical mass. In the eyes of the firm's clients—those in its target markets—the firm should have sufficient depth of capabilities and capacity to be seen as a leader, or ideally *the* leader, in each of its target markets. For example, a 20-person firm would have difficulty asserting market leadership in four distinct markets, but a 100-person firm might reasonably do so.

By applying specialized knowledge and skill to distinct but related markets, firms can leverage the knowledge from one market to gain a share in another market. In doing this, firms must take care to ensure that the marketing approach emphasizes specialized expertise in each market, which must be supported by a real commitment of resources to each market. Consider the example of a 300-person architecture firm that serves a highly diversified client base that includes developers of single-family homes, multifamily homes for sale (condominiums), and multifamily homes for rent. Although all of these markets fall under the general category of housing, they are actually very distinct markets driven by completely different demographic characteristics, building code and zoning requirements, market conditions, and their respective clients' goals. The firm has immersed itself in understanding each of these markets so intensely that it arguably knows more about each market than its developer clients. For the value embedded in this depth of knowledge, its clients compensate the firm extremely well.

Market Focus Requires Focused Marketing

A common mistake made by firms serving unrelated markets is to create a consolidated website or produce an all-purpose firm newsletter that is distributed to clients and prospective clients in all of the firm's markets. Even if your firm has highly specialized expertise and a leadership position in each of its markets, the "smorgasbord" marketing approach to those markets is confusing to your clients. Marketing activities should reassure potential clients and be designed to reinforce market focus and specialized skills, not undermine them.

If a firm accepts the premise that growth is desirable, then it has to identify growth as a major component of its vision in its strategic planning process. Then the firm needs to clearly define and identify its target markets, and the champion or champions within the firm for each of those markets—the key people who know more (or want to) and can learn more about the markets than even the firm's clients do. If clients do not perceive that people with this level of knowledge are available to them, why would they hire a design professional at all?

An equally important decision in developing a focused growth strategy is deciding what *not* to do. Every firm has a finite capacity in terms of its leaders, its staff, its marketing, and its financial wherewithal. Every hour or dollar spent pursuing—or doing—a project outside a firm's focus is an hour or dollar that is not being spent reinforcing its market position and increasing its value, the real definition of opportunity cost. Areas of weakness can consume a disproportionate share of management time that is better spent on opportunities with real promise. Once focused growth is a part of a firm's vision, the next question is how the firm should be organized—the subject of Chapter 14.

C h a p t e r

14

FIRM ORGANIZATION

Think of the way your firm is organized—its organizational structure—as a management tool that you use to leverage the talents and skills of your people, promote growth and profitability, and meet the needs of the clients and markets that you serve in the context of the geographic, economic, social, and political environment in which you operate. The way a firm is portrayed on its organizational chart reveals two important things about a firm: how it orients itself to its clients and its work, and who reports to whom within the company (though often the chart is intended to reflect only the reporting structure). In many design firms, the organization evolves organically like a crystalline structure, reflecting how relationships in the firm have developed over time, rather than as the result of thoughtful organizational design.

A firm's organizational structure affects many vital attributes of the firm: the effectiveness with which the firm brings in work; the focus it brings to clients; the efficiency with which it executes the work (contributing to its profitability); the level of collaboration, the range of career paths, and even employee satisfaction within the firm; and finally, an infrastructure for the potential growth of the firm.

As such, determining whether a firm has the optimum organizational structure is often an important element in the strategic planning process. Its organization has to facilitate the firm's ability to reach its vision, not stand as an impediment.

COMMON ORGANIZATIONAL STRUCTURES

Regardless of whether their organizational structures were intentionally designed or simply evolved over time, many design firms are organized in one of the following ways. You may see your firm reflected in one of the following descriptions.

Organizing Around Clients

When a firm is client-centered, it is organized from the outside in. Ask yourself how your clients would prefer to see your firm organized—or better yet, ask *them*. Consider what attributes bind the firm's clients together as a coherent group or groups. Take into account all of the relevant variables affecting your clients: their needs, priorities, and personalities; marketplace conditions, economics, geography, politics, the availability of labor, and the cost of doing business; and, most importantly, how they define their own success. Also bear in mind that project type—such as education, health care, or justice facilities for an architecture firm and bridges, wastewater, or geotechnical for an engineering firm—is only one facet of a firm's organization, and it may not be the most important one.

There are a number of ways to organize a client-centered firm successfully. A firm that organizes around its client-type markets can bring in specialized expertise as each project demands. For example, an architecture firm that relates to its local government clients has a target market in community buildings such as recreation centers and libraries, and could bring in a library consultant for those specialized projects that warrant it.

A firm that designs office buildings could find itself serving several very different markets. The developer of speculative offices, for example, wants to build as quickly and often as inexpensively as possible, the sooner to obtain revenue from the lease or sale of the building. A developer of civic office buildings, on the other hand, has clients that use an office building as a permanent home, and it must satisfy a much broader constituency, such as an elected executive, a legislative body, and ultimately the taxpayers. Although both developers have office buildings as their common project type, they have very different priorities.

Organizing Your Message

A 200-person A/E firm with multiple offices was neatly organized into three groups, each of which focused on a distinct project type: justice facilities (i.e., prisons), K–12 schools, and hospitals.

The firm invested heavily in marketing materials, including a high-quality, tabloid-size, full-color newsletter, which it sent to a very large mailing list of all of its clients and many prospective clients. One issue contained a cover story that featured a recently completed project, a maximum security prison for its home state's department of corrections, including a large aerial photo that made the prison look as attractive as a prison can look, including guard towers and rooftop razor wire glinting in the sunlight.

Of course, that same issue of the newsletter also went to hospital administrator clients and public school district facilities directors who may have been contemplating hiring the firm to design a kindergarten wing for an elementary school.

Unfortunately, this firm's marketing materials did not reflect its client-centered organizational structure. Despite its pride in its projects and best marketing intentions, it ran the risk of alienating valuable clients who would believe that the firm just doesn't understand them. A much better solution would have been for the firm to reflect its in-depth knowledge of its clients by publishing a separate newsletter for each of its three client groups, each of which would have communicated a singular focus on its target audience's unique needs and priorities.

Organizing Around Project Types

As stated in Chapter 13, because of the high stakes involved in the construction of any project—a home, hospital, or highway—clients understandably want to hire design firms that can demonstrate specialized experience in the kinds of projects that the clients pursue. Therefore, many firms organize themselves around the types of projects that they build. For example, an architecture/engineering firm might have one group specializing in developer office buildings and another in multifamily residential projects, each of which requires very different types of design expertise.

Organizing Around Geography

When a firm has multiple offices to serve clients or produce projects in different locations, it often makes sense to the leadership to organize around those geographic locations. If it is important to the firm's clients that the services are delivered by a firm with local presence, visibility, and knowledge, then firms often open offices where their clients or their clients' projects are located. This is a very common structure for civil engineering and surveying firms, especially those that serve residential and commercial developer clients. These firms typically place a locally knowledgeable manager in each office (and some of its offices may be in nearby or even adjacent jurisdictions) to whom all employees in that office report, either directly or through intermediaries.

Organizing Around Principals

Firms that are led by a small group of principals—who typically have known and worked with one another for a long time (often firm cofounders)—are strongly driven by those principals' personalities. Over time, the principals develop their own clientele, with whom they build solid professional relationships and cultivate good personal chemistry. For these clients, the principal is synonymous with the firm; the principal *is* the firm. As a result of relationships that have developed over time, the firm's clients may have their favorite principal, whom they hire again and again. There may be considerable overlap in the types of clients served by all of the firm's principals, but the common denominator is that each group of clients is part of each principal's personal and professional network. Often, these types of firms develop multiple small teams under each principal, each team headed by a project manager who is in day-to-day contact with the clients and develops relationships with his or her peers in the clients' organization, thereby positioning himself or herself to become a principal with his or her own group of clients in the future.

Organizing Around Services

Some multidiscipline engineering or A/E firms organize around the services they provide, with discipline-driven departments, such as an architectural design group, an architectural production group, a structural engineering group, a mechanical engineering group, an electrical engineering group, a site/civil engineering group, and a construction administration group. Each discipline comprises a department that contributes—sequentially or concurrently—to the projects and is led by a director (of design, structural engineering, and so forth). Even in such firms that have multiple offices, each employee reports to his or her department or discipline director, regardless of location, and not necessarily to the local office manager.

Organizing Around Studios

Architects often gravitate toward organizing their firms around "studios," a term applied to architecture firms for many years, also used in architecture schools, and connoting a reference to design. Like a firm that is organized around project types, the studio organization generally means that the resources to complete a certain type of project, such as a K–12 school, are found in a "studio" or department that focuses on that work. *Studio* is really a universal and nondescriptive word that may also reflect an organization by principal, client type, or geography. Therefore, a closer look may be required to determine exactly what the "studio" really represents.

The Matrix Organization

When a firm is large or diverse, it often has to reconcile how to manage a variety of work or project types, client types, and office locations. For example, a firm might have an education group, a health care group, and a corrections group, with staff in each group spread across six offices throughout the country. These firms sometimes adopt a matrix structure of organization.

The matrix structure has two lines of reporting, and the firm's organizational chart looks more like a table (a matrix) than the customary up-and-down—often pyramid-shaped—hierarchical arrangement of boxes and lines. The rows, or "vectors," of the matrix might represent project or client types, and the columns might represent the physical locations of the firm's offices, or vice versa.

An immediately evident drawback of a matrix structure is that, on paper, it appears that an employee within the matrix has two bosses. Poorly defined matrix structures can be confusing, because essentially an employee actually does have two supervisors: for example, in a multiple-office firm, the director of the water/wastewater department in a remote office as well as the manager of the local office to which that employee reports to work every day. This is confusing to the employee and counterproductive for the firm. A firm with a matrix organization is more likely to be successful if there is a dominant direction of reporting; one or the other "vectors" of the matrix must take precedence.

Imagine a piece of lumber with its grain running in one direction. It is always easier to cut with the grain than against it. Now apply this metaphor to the matrix structure: the firm must define the *dominant* direction of reporting, the direction of the "grain" of the firm's business (which, ideally, should be decided from the client's point of view, as we discussed earlier in this chapter). Alternatively, depending on the situation, the location might be the dominant vector, or the project type might be, or something else.

What Is a "Market"?

In examining how design firms are organized, we refer to terms such as *project type* and *client type*. In Chapter 13, we defined a "market" as something different: a market for design services is a group of clients who think alike, share the same concerns and priorities, and have the same goals.

Sometimes, the distinction between organizing around clients' needs versus around project types can appear fuzzy because different types of clients tend to need different types of projects, too. Rather than get too tied up with the semantics, it is always best to put yourself in the clients' shoes and study their issues to determine the primary drivers of your firm's optimum organizational structure. Although the technical skills needed to design and produce different project elements may be very similar—foundation design, wall sections, and roof details—and may even be transferable from one project type to another, the approach to each project type (and client) is quite different. Design firms should be organized from the outside in, from the clients' point of view and not for the convenience of the managers of the firm. Asking yourself why clients hire your firm—or any firm—will yield information that will guide your organizational structure.

Organizational Overlaps

A firm that organizes around its client-type markets can also take on the form of other organizational structures that follow geography or project type. For example, the clients of a site/civil engineering firm, though they may focus on many different markets such as residential development, shopping centers, or sports facilities, all have in common that they value local site/civil engineering expertise—not only knowledge of the local geologic and hydrologic conditions, but also knowledge of the regulatory environment, how to work with local authorities, and how to obtain their all-important entitlements and permits. From the client's perspective, an engineering firm that understands the nuances of the highly localized regulatory approval process which varies from jurisdiction to jurisdiction, and that has developed strong relationships with the local authorities can save its clients an enormous amount of time, and therefore money. Especially with the advent of technology, the engineering firm could easily import its technical skills—grading sites and designing curb and gutter schemes—from anywhere in the country or the world, but its real value to its developer clients that transcends its technical expertise is its local presence and ability to secure permits. Therefore, in this scenario, the client-centered site/civil firm responds to its clients' priorities by organizing around local geography.

ORGANIZATIONAL STRUCTURE AND STRATEGIC PLANNING

At some point, a firm's organizational structure may no longer effectively support its operations, or it is not aligned with the firm's mission or its leadership's vision of what the firm can achieve in the future. When that happens, organizational structure becomes an important issue in a firm's strategic planning process.

Of all the issues that a firm tackles in strategic planning, changing its organizational structure can have a profound and pervasive impact on its people, operations, business processes, and finances of the firm, as well as on the way clients and potential clients perceive and interact with the firm. The daunting prospect of reorganizing a firm must meet the test of any strategic planning issue: either the firm must change the structure in order to achieve an even greater level of success than it is currently achieving, or the existing structure is an obstacle that impedes the firm's ability to be successful.

Considering the variety of design firm organizational structures described above, there is no one right or wrong way to organize a firm. There are, however, a few proven guiding principles for determining and designing a firm's organization:

1. The organization must respond to the external marketplace and meet the needs of clients. This seems obvious at a glance, yet, as we have stated, many design firms organize for their own convenience or out of years of habit without deeply examining whether their firms meet this first crucial test.
2. The organization must operate effectively. Although all designers like to think that they are providing added value to their clients, the vast majority of firms are in the business of selling time—billable hours—and the organization must contribute to the efficient use of those hours.
3. The organization must take advantage of the talent within the firm, leveraging people's capabilities and providing them clear paths for career development and advancement.

DESIGNING YOUR FIRM'S ORGANIZATIONAL STRUCTURE

Changing a firm's organizational structure is not a strategy to be taken lightly and can be one of the most sweeping and challenging aspects of a firm's strategic plan. After all, changing the way a firm is organized may alter every aspect of the experience of working with and within that firm. The change is bound to push some managers and employees beyond their comfort zones. When the changes are positive, as they should be if a firm makes them as a result of a thoughtful strategic planning process, they are still bound to breed anxiety.

Resistance to Change

Even if the shortcomings of the firm's existing organizational structure are widely acknowledged, people are resistant to change; they prefer to do things the way they're used to doing them, even if that way is unnecessarily tortuous. Often, when a firm's organizational structure is not effective, people nevertheless learn how to work within it, in terms of their roles and responsibilities, to whom they report, the kinds of work they are supposed to do, and how they fit into the firm overall. However, if the most effective firms are managed from the top down, then it doesn't make sense to allow the organization to be managed reactively, via "work arounds."

A new organizational structure can seem particularly threatening to people in supervisory positions. This is because when a firm is reorganized, people may perceive the changes in terms of personally winning or losing. Take the example of a firm that is organized geographically, with each local office led by a principal responsible for the performance of that office. If the firm decides on a strategy to reorganize according to the client-type markets it serves, the office-manager principals may suddenly find that they no longer have an office full of employees reporting to them. The empires they have built will have been taken from them. However, they still have great—perhaps even greater—value to the firm. Some may become market leaders—principals in charge of a particular client-driven market segment—while others may perceive themselves as having been demoted to senior project manager. In the face of major organizational changes, it is vital for the firm's leadership to communicate clearly and convincingly to everyone in the firm that the changes are for the greater good of the firm, and that the changes in supervisory, budgetary, or financial performance authority do not reflect negatively on the ability or performance of any individual.

Starting With a Clean Slate

If the strategic planning process determines that your firm needs some degree of reorganization, the planning team should begin with a clean piece of paper rather than be burdened by legacy issues of the firm's current organization. Although it is inherently very difficult to accomplish this without considering the people in your firm, the most effective way is to set aside personalities and internal politics and resist the temptation to put names in boxes. By first designing your firm's organization the way it would work best, you can then assess who are the best players for each position and where you may have to bring people up from within or even from outside the firm to make the new organization sing.

As with any change that your firm makes in strategic planning, remember that a new organizational structure can't really be imposed upon a firm. The people responsible for implementing it have to support it and believe in it in order to make it work, and they must communicate that support so that it permeates the rest of the firm. Broad-based buy-in is crucial to the success of a reorganization.

Aligning Performance and Rewards With the Organization

A strong indicator of how a firm is really organized—if it's not evident from the organizational chart—is how the firm measures its financial performance: around what business units it monitors, manages, and incentivizes its profit and performance. If the firm measures profits and losses by office, then it is organized by office location. If the firm measures the performance of the wastewater facilities group separately from the transportation engineering group or the hospital group, for example, then it is organized by professional division or service.

The way in which a firm's organizational structure incentivizes and rewards people is a matter of some controversy among design firms.

- Some firms employ incentive compensation programs that directly reward managers and their employees for the performance of their business unit—again, defined by the organizational structure as a market group, office, etc. When their group performs well, they share in the profits generated by their group. When their group does poorly in terms of financial performance, the people in the group do not receive rewards. The primary argument against this philosophy is that it focuses managers' and employees' attention too narrowly on their own group's performance—sometimes at the expense of other groups—and stands in the way of firm-wide collaboration.

• Other design firms have a Three Musketeers "one for all and all for one" incentive compensation program in which everyone in the firm—regardless of their home group—is rewarded from the same firm-wide profit pool, sometimes called a "one-firm firm." This is intended to encourage collaboration across groups and capitalize on the entire firm's resources, but the main argument against it is that one poorly performing group can drag down and penalize everyone in the firm.

As with the organizational structure itself, a firm should determine the best incentive and reward system for itself consistent with its long-term vision and its strategic plan.

15

FIRM MANAGEMENT

The old saw in the world of design firms to "get the work, do the work, and get paid for the work" is simplistic but valid, and a strategic plan should address all three of these vital functions. But there are distinct issues pertaining to how an entire firm is led and managed that can be separated from other operational matters that more directly affect winning, executing, and delivering work to clients. Issues related to managing a firm can be among the most thorny that a firm will face and the most difficult to address in the strategic planning process because they are related to the performance or conduct of senior members of the firm—perhaps even the most senior leader of the firm. The word *accountability* arises frequently and consistently within design firms embarking on strategic planning, an indicator of how ubiquitous and troublesome this topic is among architecture, engineering, and planning firms. More than any other issues, those related to corporate management must be addressed openly, with honesty and courage.

ESTABLISHING A CLEAR LEADERSHIP MODEL

Singular leadership of a firm is far preferable to management by committee. Some*one* must be responsible for leading the firm in fulfilling its mission and achieving its vision. When this responsibility is distributed it becomes diluted, no matter how well a group of equals works together. The advantage of singular leadership is its simplicity and clarity: everyone understands who's in charge; no explanation is required.

Management by committee, by contrast, is inherently ambiguous and unwieldy. Even in the best of circumstances, consensus decision making can become a slow and cumbersome process. Such an organizational structure is rarely the product of thoughtful design. More often, it is the default arrangement of second-generation firms where a strong founder has failed to designate and groom a successor, and the members of the second tier cannot bring themselves to elevate one of their own to lead them. This second-generation scenario is so common that one of the primary responsibilities of a firm leader should be identifying, designating, and mentoring a single successor (see Chapter 16). Founding firm leaders especially often have great difficulty designating their successors, because identifying and designating future leader *A* means not promoting persons *B*, *C*, or *D*. The concern is often unwarranted, because the same leadership skills that a conscientious retiring leader recognizes in an individual are likely to be apparent to everyone, including the nonselected *B*, *C*, and *D*. If the choice is made with sound judgment and in the best interest of the firm, most staff members will accept, respect, and even applaud the retiring leader's decision.

The tendency for design professionals to drift into management by committee may be an outgrowth of the collaborative nature of design, or a reflection of the types of people drawn to the design professions. Most architects, engineers, and planners are, in fact, collaborative and nonconfrontational people. They choose their profession because they have a passion for it, not because they have a drive to build an empire, or an all-consuming ambition for financial success. They often tend to think of their firms not as typical business enterprises, but as unique working environments that they have created and nurtured. Although conventional business wisdom and the lessons learned by many firms overwhelmingly suggest that management by committee simply doesn't work, many firm principals rationalize that conventional wisdom does not apply to their own firm and its culture. This belief is a fallacy that can threaten the long-term viability of the firm. Design firms are design *businesses* that need to be run like businesses if they are to survive and thrive.

If your firm is mired in management by committee, you can consider the following steps to move toward more effective singular firm leadership:

1. The members of the committee have to subordinate their own egos to the greater good of the firm. If they cannot accomplish this, the firm will be doomed to a lack of bold decision making for the life of the committee.
2. The committee members should develop a list of criteria they would like to see in their firm leader.
3. The committee members should individually objectively score each other—and perhaps other potential candidates within the firm but outside the group—against these criteria.
4. The committee members should share the scores and candidly discuss who is best suited to lead the group and, by extension, the firm.

MANAGEMENT TITLES VERSUS ROLES

Many design firms function as real or de facto partnerships, with multiple individuals sharing the title of principal, even if the firm is formally incorporated as a Subchapter S Corporation or C Corporation for legal and accounting purposes. But "principal" is more of an honorific than a functional job title and does not delineate a specific set of responsibilities. Usually, the title of principal is either a direct indication of ownership status or is more generally intended to communicate to clients that they are dealing with a senior member of the firm. It is a meaningful indication of the individual's stature and general level of responsibility within the firm; no firm would name an individual principal unless it believed that the person exemplified the qualities that the firm wishes to exhibit in the marketplace.

Similarly, the nonfunctional honorific title "associate" can serve as an indication that the firm has confidence in the individual and that he or she has a bright future with the firm. It means that the person has achieved a certain level of responsibility for which the firm wishes to confer recognition. In many firms, associates form the pool from which principals are selected.

In addition to the added responsibility of the title, among the criteria for elevation to principal or associate should be whether the current owners wish to welcome that person as an owner now or in the future. If a person is named associate or principal, that person should want to invest in the firm, literally with their cash and figuratively through their work. Correspondingly, at some point in the process and depending on the firm's vision for its ownership structure, the firm should *expect* its most valuable contributors to willingly invest in the firm. People who want a firm to stake its future on them should be willing to stake their future on the firm in a reciprocal transaction and mutual expression of trust and confidence. Chapter 17 discusses ownership in more detail.

The exact meaning of the titles of principal and associate are often questioned, especially by junior employees who have no clear concept of the titles' significance. It is difficult for them to envision climbing the career ladder when they cannot see the rungs. Some dynamic and energetic associates contribute more to the success of their firms than some "coasting" principals, which further muddles the meaning of the titles and breeds resentment. A design firm should define precisely what these titles mean, and what the implications are of being named an associate or principal. It is counterproductive to reward employees with titles that only serve to confuse, or to perpetuate titles for individuals who do not measure up to the performance expectations commensurate with their titles.

The lines of reporting can become very blurry in design firms because some senior members of the firm may have multiple roles all at once: employee at will, principal or associate, shareholder (owner), and perhaps even member of the board of directors. It is very important for everyone to understand and accept that governance, leadership, management, employment, and ownership are all completely different and separate concepts, and that each role comes with distinct responsibilities and rewards. A principal who is also a board member remains accountable to the chief executive officer for his or her job performance, but would separately exercise his or her fiduciary responsibility to the firm as a board member elected by the shareholders.

LEADERSHIP AND GOVERNANCE

There are two levels, or platforms, of firm management. *Governance* is the umbrella of overall firm management, regardless of the firm's organizational structure, that provides accountability to the firm's shareholders and provides a mechanism for long-term strategic decision making. Everything above the firm leader—president, CEO, or managing principal—is a matter of governance and everything below is a matter of operational management and organizational structure. Governance authority is usually invested in a board of directors, representing the interests of the shareholders, to which the CEO is accountable for the firm's performance. A firm that does not have a fully functional board of directors is missing an essential element of accountability at the highest level. Even smaller firms need effective oversight and a strategic decision-making process, in part to protect the owners having a minority ownership interest, but also to provide the firm leader with a structured group of advisors, each of whom understands and exercises his or her firm governance responsibilities.

Design firms, especially those that pride themselves on an informal culture, are often not accustomed to—or even comfortable with—operating in a corporate fashion with true levels of accountability. It makes their founders or firm leader(s) feel as if the firm doesn't trust them to do their jobs. It also seems as if unnecessary bureaucracy will get in the way of their decision making. In fact, some design firms use the word *corporate* pejoratively, as in "we don't want to get too corporate."

In design firms with more than one owner, a composition that constitutes the great majority of firms, a common scenario is that of a single individual who maintains a controlling ownership interest in the firm in order to manage it as he or she pleases. In this situation, there is relatively little in the way of accountability to the other shareholders. Founders of firms sometimes share this strong tendency to retain majority ownership and management control of their firms. They believe that they have earned what they have built and that they have the right to the rewards. But this is never entirely true. A founder of a design firm cannot build value without the collective effort and dedication of many people, especially the other shareholders.

LEADERSHIP VERSUS MANAGEMENT

Whether the firm leader has the title of CEO, president, managing principal, or managing partner, his or her primary responsibility is to maintain a long-term focus on the future and guide the firm toward it.

Put more succinctly, a CEO's job is to develop the firm's strategic plan. Everything the CEO does should be directly related to that end. The CEO continually inspires, motivates, and cajoles the firm's managers and its employees toward achieving the firm's vision.

It takes a certain quality of character to be an effective design firm leader. Leadership and management are not the same thing. Although the size of a firm has a bearing on the leader's direct involvement with clients or in projects, many design firm leaders—even of larger firms—get entangled all too easily in the minutiae of running their firms or the firms' projects. This is a common inclination among design professionals, who are comfortable managing the professional and technical issues for which they have been trained and educated, and to whom project problem solving comes naturally. But a micromanaging leader who becomes too deeply involved in the day-to-day issues of running the office or getting projects done reverts to becoming little more than a highly paid project manager. If the captain of a ship is preoccupied with shoveling coal into the boiler, who is navigating and steering the ship? Managers make sure the boilers are stoked; leaders see the ship's destination and pilot it safely there.

The temptation to drift away from the constant and frequently lonely challenges of leadership and toward more comfortable client and project management responsibilities is particularly acute for the leaders of smaller firms who must, of necessity, devote some of their time to management responsibilities and project involvement. After all, small firms succeed in part because their principals remain close to clients and responsible for projects. But even the leaders of smaller firms need to prioritize their responsibilities by asking themselves, "What is my highest and best use?" and continually remind themselves that leading the firm today and grooming a leadership successor is their top priority, even in the face of their many other day-to-day responsibilities.

The Attributes of a Leader

In addition to the qualities that we discuss in this book that a design firm leader should exhibit and embody in the strategic planning process, there are other, more general attributes that managers and leaders at all levels in a firm, including the design firm leader, should aspire to exemplify.

It's a short list to remember, but a tall order to live up to. An effective and respected leader:

- Is authentic, sincere, and forthright
- Behaves predictably
- Communicates strongly, clearly, and consistently
- Is a role model and inspiration to others
- Personifies the values of the firm that he or she leads
- Displays an energetic and conspicuously strong work ethic
- Radiates energy and passion for the success of both the firm and the individuals within it

Part of being a good leader is recognizing the difference between management and leadership.

MANAGEMENT VERSUS LEADERSHIP

Operational management, as opposed to firm leadership, is concerned with coordinating and improving the functional components of the business that support the ability to generate revenue and profit, including the following:

- **Project-related management**: overseeing the work of project managers and their teams (but not managing the individual projects themselves) to ensure that the execution of the work is acceptable, that projects are properly staffed, that project teams are providing a satisfactory level of service to clients, that the quality of the work is being maintained, and that projects are being produced profitably and on schedule
- **Human resources management**: includes recruiting, staff development and training, and staff retention strategies
- **Financial management and accounting**
- **Information technology**

If a firm is large or complex enough, overall management of the firm's operations may require a chief operating officer (COO), someone who manages the business and internal project operations of the firm. A crude rule of thumb is that when a firm has more than 150 to 200 employees and/or multiple offices, it needs (and can support the cost of) a COO. For firms up to a certain size and organizational complexity, one person can function as both CEO and COO, provided that person has the ability to distinguish and maintain the proper balance and separation between the two roles. Also, as firms initially recognize the need for higher level operational management and coordination, the newly created position of COO may not require a full-time commitment at first.

Because design firms tend to promote from within—in part because of the risk associated with bringing in senior people from outside— most firm leaders step up to the firm's leadership role from a senior management position. Considering the differences between management and leadership, evaluating the ability of a prospective successor to make the transition from manager to leader is crucial to the future success of the firm.

DISPELLING THE MYTH OF TENURE

There is no meaningful correlation between an employee's value to a firm and his or her longevity with the firm. New employees with experience in other firms or even other industries can bring fresh ideas and make significant contributions to a firm. Conversely, many employees who may have been with a single firm for most or all of their careers can become dangerously narrow-minded and limited in their contributions if they do not consciously expand their worldview. In considering candidates for leadership positions, firms often display a bias toward senior people who have a longer tenure with the firm. It stems from a shared sense of entitlement, a belief that "it's his or her turn;" that a leadership post is the proper reward for many years of service. However, someone younger or with a shorter tenure may be more qualified and devote more creative energy to meeting the challenges of the top spot.

Attributes of leadership are not necessarily age-related, but many firms find it difficult to violate an unspoken social and business protocol by leapfrogging over long-tenured employees to select a younger person as the next firm leader, or even bringing in a new leader from outside the firm. The only way to counter that tendency is to focus not on the welfare of an apparently deserving individual, but instead on the future success of the entire firm upon which all shareholders and employees depend for their career, investment, and livelihood. Firms should find ways to reward or recognize people for many years of service other than to put them into positions in which they will underserve the firm.

MANAGEMENT PERFORMANCE AND ACCOUNTABILITY

When leadership succession is, or should be, on the horizon, it can become the most critical and sensitive of strategic planning issues, but poor—or more insidiously, lackluster—performance by managers runs a close second. The higher the management level of the individual concerned, the more sensitive the issue is and the more likely it is to be sidestepped due to aversion to confrontation on the part of the CEO and the manager's peers. The senior management of every firm has to agree on an important fundamental precept in order for the firm to be able to address sensitive performance issues consistently, promptly, and forthrightly: if everyone in the firm is mutually dependent on the performance of everyone else in the firm, then everyone has to report to—and be accountable to—someone. And if everyone is accountable to someone for their performance, including associates, principals, and even the CEO, then rationally and objectively addressing performance issues is simply a matter of treating a manager's performance like any other employee performance issue.

Whether an individual is a division head, a department head, or a project manager, someone is responsible for coaching, regularly evaluating, addressing the performance of, recommending promotion of, and (if necessary) disciplining or even terminating that person. The same accountability must apply to principals. Regardless of their status as shareholders, principals, like all other employees, need to be accountable to someone for their performance, which should be measured as much as possible with objective performance standards. When clear lines of decision-making authority are lacking and principals do not report to anyone—or feel that they should not have to report to anyone—above them in the firm's hierarchy, it can be nearly impossible to hold principals accountable for what they do and do not do. Considering that principals are the most highly compensated and visible people in a firm, a lack of individual accountability is especially egregious. The margin of profit in most design firms is too thin to be able to carry low performers, especially highly paid, subpar performers. Similarly, the president or CEO should be accountable to the owners of the firm for the firm's performance by reporting to the shareholder-elected board of directors.

Hanging on to Ownership for Life

As one president of a 50-person engineering firm frankly and privately admitted, "I'm very uncomfortable selling my shares, because once my ownership drops below 51 percent, the others could vote me right out of here." His concern may have been exaggerated—or not—but it revealed an embarrassing grain of truth. Had this founder and still majority owner felt more secure in—and accountable for—his performance over the years, some of his actions or inactions as the firm's leader might have been called into question, and he could have explained them on their merits. Instead, his continuing death-grip on majority ownership served as his primary job security.

The performance of many managers at various levels within a firm can be measured, but by what metrics can the performance of a CEO be evaluated? At the core, the CEO is responsible for the entire firm's performance, so revenue and profit are the most obvious. But there are many financial and nonfinancial performance metrics that contribute to those two: marketing success, recruiting and retention rates, project management as measured by budgets and cost overruns euphemistically described as "unbilled direct labor," billing and collections, and managing cash flow and debt. In short, the CEO of a design firm is no different from the CEO of any company, whether publicly held or privately owned. He or she is responsible for returning value to the firm's shareholders (most often including himself or herself).

From time to time, every firm has its "coasters," as management consultant David Maister calls them, senior people who may be highly skilled or talented but who exert only modest effort. The level of tolerance within some design firms for such individuals borders on the remarkable. Some managers who feel they have earned their lofty positions become so highly skilled at doing their jobs minimally that colleagues and supervisors cease to have reasonable performance expectations of them. They are allowed to coast along in a way that would not be tolerated in other, more junior-level staff. Allowing people to coast destroys morale. It is particularly corrosive when the coaster is a firm principal. The responsibility to impel these individuals to a higher level of performance or remove them from the firm cannot be ignored by the firm's leadership. Bright, young, ambitious associates will look at the inequity of the situation and ask themselves (or worse yet, ask others in the firm), "Why should I work so hard? What's in it for me?"

If a firm articulates its core values and is willing to include accountability among them, invoking violations of those values is the objective way to address these "sensitive management issues," which is really a euphemism for a senior manager's compensation exceeding his or her performance. These issues are seldom clear-cut, particularly when one is evaluating the performance of a firm principal. It is easier to confront these types of issues when objective metrics are used to measure the performance of components of the firm. For example, a principal might be leading a studio that is not bringing in the expected amount of revenue or profit. There may be some external reasons for the poor performance, but perhaps that principal is not putting in as much effort as other principals, or has a management style that fosters poor morale and high turnover. Only by having such metrics can a firm judge the performance of managers objectively and bring such issues to their rational conclusion.

16

LEADERSHIP SUCCESSION AND CONTINUITY

Sooner or later, it happens to every company: for any number of reasons, the leader of the firm—CEO, president, or managing principal—departs. The separation may be due to retirement, or it may be involuntary. Regardless of the reason for the firm leader's departure, the succession of a new leader must follow. Leadership succession, which is inevitable, has the potential to exert a major impact on a firm's reputation, operations, culture, and even survival; however, despite its importance, a great many—if not most—design firms don't adequately prepare for it.

Of all the many attributes that successful design firms share, effective leadership is the most important. If a firm needs capital, it can borrow it; if business is slow, a firm can boost its marketing program. But without an effective leader—today or in the future—a firm faces serious trouble.

When a design firm anticipates the inevitable transition of its leadership and plans, manages, and implements the succession process well, it leaves that firm in a good position for the future. Where many firms fall down on the job is in identifying a successor to the firm leader well in advance of the anticipated date of the actual transition. This lack of planning can severely destabilize a firm by sparking concern among employees and clients, starting rumors, and increasing overall turnover and defections of the senior managers. In a worst-case scenario, it can even result in the failure of a once-thriving firm.

The longer a leader waits to identify and groom his or her successor, the larger the apparent leadership gap looms and the fewer the options open to fill it. As a result, firms experience a sort of reverse Darwinism: while a hesitant leader agonizes over who, if anyone, is capable of filling his or her shoes, the most capable, ambitious, and impatient of the potential future leaders leave to join competitors or start their own practices, and those less ambitious who are content to wait it out ultimately take over. Instead of "survival of the fittest," these firms experience a "departure of the smartest."

Leaving a Legacy Is Important

At the outset of opening their businesses, firm founders may not initially conceive of their firms as having a lifespan longer than their own careers. But at some point, most first-generation founders and all second-generation principals become aware that the firm has a life of its own and feel a desire, or at least an obligation, to perpetuate it. The greatest legacy these leaders can bestow on their firms is to identify and groom a qualified successor.

PLANNING FOR LEADERSHIP SUCCESSION

A design firm of any size can have multiple leadership positions that it will someday have to fill in addition to the top position, whether or not the circumstances and timing of the transitions are anticipated. Even under the best of circumstances, leadership transitions are disruptive to a firm. Successful firms keep the leadership pipeline filled by grooming successors for key positions so that the firm is prepared to fill the leadership positions that it expects to open up—as well as those that it doesn't expect.

A leadership succession plan also benefits a firm in other ways. When they are recognized in advance of the transition, the designated successors understand that their supervisors see them as having the potential to advance to a position of greater leadership responsibility, a desirable career path for many design professionals. This advance knowledge and vote of confidence motivates key people to stay with the firm and to continue to contribute to its success.

Identifying potential successors in advance also has a positive effect on the rest of the staff: it is a clear signal that the "heirs apparent" are highly regarded by the senior management of the firm, and it gives the staff time to adjust to the idea that these individuals will be moving into positions of greater authority in the future. It sets up the successors as role models, illustrating what it takes to advance professionally in the firm. As we describe in Chapter 15, in most cases, when successors are chosen wisely and well in advance, their coworkers recognize—or will come to recognize—the same leadership qualities in them that are apparent to the firm leaders and will readily accept them as future leaders.

Leadership Succession Versus Leadership Continuity

The conventional terms for a change at the top used within the design professions, and therefore often used in this chapter, have been *leadership transition* or *succession*. Transition implies a change. When a firm has problems—especially related to its leadership—it is in need of leadership transition or succession, a change in course for the better. However, when a firm has been successful—due in some part to its effective leadership—it still may be undergoing a leadership succession. Successful, strategic firms think in terms of leadership *continuity* rather than *transition*. Although the difference may appear to be simply semantic, this is a way of thinking more constructively and positively about leadership succession.

CHOOSING A NEW FIRM LEADER

Of all the major issues that serve as catalysts to prompt firms to engage in strategic planning, the pending succession from one firm leader to the next is among the most common. The issue may be raised by the current and soon-to-be retiring leader or by any of the other key managers in the firm, who presumably compose the next generation of leaders, and from among whom a successor to the firm leader may be chosen. For the outgoing firm leader, the path to choosing a successor can be thorny. A firm leader planning to retire may not see an obvious successor among the next generation of management, indicating the need for more intensive training and grooming of those with potential, or perhaps even a strategic hire. On the other hand, there may be several suitable candidates, leaving the leader unsure of whom to choose. It follows that if several people have been promoted to key management positions in a firm, all of them make some valuable contribution in that role. Therefore, whether there are many capable candidates or more aspiring candidates than capable candidates, the firm leader will not want to alienate those whose aspirations exceed their capabilities; just because a key employee may not be ready to take over the firm, he or she may still be valuable and important to keep.

Members of the succeeding generation—those who feel that they are poised to take on a higher level of responsibility and potentially be selected to run the firm—will also have their own ideas about how to handle the leadership succession. Even if—or especially if—the firm has not engaged in formal discussions about leadership succession, the employees will no doubt have discussed the subject among themselves; after all, the succession will have a significant impact on the firm and on their careers.

Preparing for Leadership Succession: Guidelines for the Firm Leader

Is your firm prepared for leadership transition? Have you cultivated the next generation of leadership? Do you even have potential leaders to cultivate? If you are the leader of a firm, answer these questions:

- How do you envision your firm continuing—and improving—beyond your own tenure?
- Do you tend to think of yourself as indispensable? If so, why?
- Is the next generation of leaders in your firm—supposing there is a next generation of leaders—better than you? Are they smarter, faster, more intuitive, more entrepreneurial? If not, why not?

If you're not fully satisfied with your answers to these questions, consider implementing the following steps toward leadership continuity planning:

- Wherever you are in your career, take a broad and open-minded look at the capabilities of others within your firm. Solicit input from others if it helps to stretch your own impressions of people.
- Search for up-and-comers who exhibit signs of leadership talent and help them to develop their abilities. Remember, someone once gave you a chance to show what you could do.
- Let your potential future leaders show their stuff by having them help to develop your firm's long-term vision and take part in implementing the strategic plan.
- Begin to share, and then eventually delegate, some of your own leadership authority. Let the future leaders take on more until they ultimately take over. That's the way to achieve not just leadership transition from one generation to the next, but also true leadership continuity for your firm.

THE FIRST TRANSITION: FROM FOUNDER TO SUCCESSOR

Many design firms are only today facing their first leadership transition: from the founder of the firm to the next generation. The daunting prospect of leadership transition from a firm's founder to the second generation is often the tipping point that pushes the development of a design firm's very first strategic plan.

Design firm founders may find it difficult to plan for their own succession, for very understandable reasons. Their failure to plan has less to do with denying their own mortality than with a deeply held belief that no one could possibly do their job as well as they can: after all, no one in their firm ever has.

Founders often keep things very close to the vest when it comes to their company. Based on their early experiences, they sometimes believe that they can manage the firm single-handedly ("Nobody else understands what it takes to run this place!"), see no need to share financial or other business information ("They wouldn't understand how our finances work!"), negotiate the firm's contracts ("I've got to protect my own skin!"), and are the only ones able to personally maintain the firm's most crucial client relationships ("I go back a lot of years with these people!").

When the founder has not shared his or her knowledge or relationships with subordinates, the second tier of management may end up consisting entirely of "super" project managers, people who have developed considerable skill at managing the project process but who have had little opportunity to lead other aspects of the firm's management or business. There may not be a competent successor among the firm's next generation of leaders because the current leader has purposely—or more likely inadvertently—sheltered them and consequently constrained their development into potential future leaders.

Founders may also view their key subordinates in a paternal way. They see their potential successors as the young professionals fresh out of school that the employees once were, embarking on their careers in the founder's beloved field, and under his or her wing. This natural tendency to remember things as they were as opposed to seeing things as they are can impede leadership succession—not to mention the career development of those who otherwise would be rising stars. It may be hard for the firm leader to see the younger generation as professionally mature and prepared to step up to a new role and position in the firm, even if their colleagues and clients perceive them as ready to do so.

If a firm does not have clearly qualified candidates for succession, it may be an unintended consequence of other circumstances that founders sometimes foment. By their nature and by necessity, founders often are gregarious and extroverted, possessed of an entrepreneurial spirit and a high tolerance for risk; these are key attributes for starting and growing a successful business. Their position and personality traits have enabled them to develop a broad and deep network of professional connections with peers with whom they trade intelligence and share knowledge. To the succeeding generation, these attributes may seem inborn—which they may be—and not acquirable, though they generally are. It is common for those in the second generation to share the belief that the firm leader is simply irreplaceable, with shoes too large to fill.

If the next generation has been deprived of proper mentoring and training, they very well may *be* inadequate, at least for the time being. If the firm leader has excluded staff members from business decisions, finances, and contract negotiations, for example, they can't very well be expected to step naturally into a leadership role, much less have any confidence in their own management skills and leadership abilities. One way that firms address this situation in their strategic plans is to develop and institutionalize a leadership mentoring and training program for promising employees, and by being open with employees about the firm's business practices and financial performance.

When the Heir Isn't Apparent

Sometimes, no one in the next generation displays even the raw materials from which to fashion a future leader of the firm. In these cases, the outgoing leader's options are more limited:

- Choose an individual from the next tier who comes *closest* to having "what it takes" and stay in the leadership position longer than originally planned to belatedly begin a mentoring process.
- Look for a leader outside the firm, most likely from a competitor or a teaming partner.
- Search out another firm to acquire, and in the process acquire that firm's next leader.
- Find another firm to acquire your firm.

Faced with these alternatives, some firms choose to bring in the next generation of leadership from outside through a strategic hire. Although there are success stories of firms who have played out this scenario, it does carry with it a greater degree of risk. Every firm has its own culture, and an outsider—especially at the very top of the organization—may not "fit." Also, any new leader, just like any new hire, irrespective of the amount of due diligence performed, comes with "baggage," those unknowns regarding the person's capabilities, personality, and even work habits. A candidate with the strongest résumé and best showing at the interview may still prove to be a failure on the job. "Undoing" any hire is awkward, but reversing the hiring of a new firm leader may call into question the judgment of the outgoing leader.

One of the best approaches to bringing in a new leader from outside is to hire that individual into a senior leadership position first for at least one or two years as a "proving ground" before taking the top spot. For firms large or complex enough to need a chief operating officer (COO) or director of operations, that role—while requiring a somewhat different skill set than that of the firm leader—can at least serve as a meaningful test prior to a complete commitment.

PREPARING THE STAFF FOR LEADERSHIP SUCCESSION

Naming a successor to the firm leader is always a matter that firms must handle diplomatically, but not hesitatingly; this is a time for the outgoing leader to call up professionalism and empathy, and to handle the situation with care and candor. Other members of the firm may be anxious about how their work life will change under the new leader's direction. In smaller firms, most of the staff has the benefit of knowing the new leader, but if the firm is large enough, many people will know the new leader only from reputation and rumor.

When a new firm leader is designated, the rest of his or her generation will have to accept that one of their peers will move ahead of them to become their organizational superior. This is naturally very difficult for even the most mature, self-assured person to accept; even if the subordinate views the new firm leader as worthy of the position, the harsh fact remains that the former peer has been selected to lead the firm—and the subordinate has not. If the new leader has been with the firm for a shorter time than the subordinates, this may serve as salt in the wound.

An awareness of this uncomfortable situation, which transpires even when all of the group of future leaders from whom the designate is chosen have healthy working relationships and mutual respect, prompts some firm founders to abdicate their responsibility and leave the succession decision to the second generation, in effect tasking them with choosing a leader from among themselves. Unfortunately, this is more likely to be perceived by the staff as indecisiveness on the part of the firm leader rather than egalitarianism, and it tends to increase any anxiety about the transition among the management group and among the staff. The retiring leader may regard the gesture of delegating the decision as magnanimous and even wise, believing that he or she should not impose his or her will on others beyond his or her tenure. But the underlying message to both the successor generation and the firm's entire staff is that the outgoing leader is uncertain—certainly not a ringing endorsement of anyone in the next leadership group.

An Unsuccessful Succession

The leader of a 120-person A/E firm recently retired. Far in advance of retirement, he had identified six second-generation principals who would succeed him as owners of the firm when he left. Among that group, one stood out for his leadership ability, but the founder never formally designated him as the heir apparent out of fear of the reaction from the other five. All six individuals were highly skilled in project management, client relations, and even studio management, but only this particular individual had the strong business savvy needed to lead the firm. Upon the retirement of the founder, the six principals could not bring themselves to name the best-qualified leader as president, and in so doing, agree to report and be accountable to him. The firm muddled through with management by committee for five years—achieving below-industry-average performance—before eventually and reluctantly elevating the person to "managing principal." However, the other five principals could not truly accept his leadership, even in this role. They found him too demanding. Their clear line of reporting to him grated against their pride, so ultimately they fired him.

The outcome might have been different had the founding principal named the obvious leader as his successor. One or more of the other principals might have voluntarily left the firm as fallout from the decision, but the overall outcome for the firm would likely have been more positive. Decisions would have been made more quickly and effectively, and the firm would have progressed more assertively. Under their management-by-committee structure, consensus was just too difficult to reach and sound decision making impossible to achieve, particularly on difficult issues.

OBSTACLES TO LEADERSHIP SUCCESSION

When the Leader *Is* the Firm

One of the reasons that leadership succession is so difficult to address is that it involves the personal feelings of the highest-level people in a firm. Leaders can identify themselves so closely with their firm that they cannot imagine anyone disputing their ability—or their right—to make unilateral decisions about anything and everything. Firms—especially smaller firms—can develop a personality or culture that mirrors the personality of the leader, and when this happens, it can be very difficult to find a successor who will continue and perpetuate that firm's personality.

If a firm leader has certain undesirable personality traits, or ineffective or counterproductive management approaches, the subordinates may figure out ways—often ingenious—of working around the leader rather than confronting the personal leadership issues directly. Firms can and do limp along that way for a very long time, but they are unlikely to thrive, especially in a marketplace that is highly competitive for clients and for staff. See Chapter 15 for a more detailed discussion of firm management models.

When the Wealth Isn't Shared

A second common obstacle to leadership continuity has to do with compensation. When a firm has been financially successful over time, the share of the profits that accrues to its owners, through base salary, bonuses, and especially ownership distributions, can increase much more quickly than the compensation of the nonowner employees. The owners' total compensation including salary and profit distributions—especially if the ownership is concentrated among a relatively few individuals—may become so large that retiring firm leaders actually feel embarrassed and become reluctant to reveal it to their successors.

Of course, owners are entitled to earn as much income from their firms as they wish. However, out-of-balance compensation tends to occur in firms in which the ownership is concentrated in a single owner or a very small group of owners. The firm may be performing so well that it is able to compensate staff members fairly, even generously, but there remains so much profit that, when it is distributed among a small group of owners, the individual sums are considerable. Often, such firm owners are not deliberately trying to rake as much as they can from across the top; it just happens over time. When leadership succession becomes necessary, the moment when the compensation arrangement is revealed to the leader's successor can be uncomfortable for the retiring owner. To avoid this situation, prospective future leaders should be exposed to and understand the firm's finances, not be shielded from that information.

Sharing the Wealth

A 120-person A/E firm was owned by just two individuals. The firm had been extremely profitable for years, with its employees enjoying good salaries and generous bonuses. There was so much profit accruing to the two owners, however, that they felt that they couldn't allow anyone in the firm to know how profitable the firm really was, because they were concerned that it would have actually reflected poorly—rather than favorably—on them personally. They honestly feared a professional mutiny. As they each approached their respective retirements, the two owners identified four people to take over the management of the firm. One would be president, another CFO, a third would head operations, and a fourth would be the lead designer. Of the four, only the CFO-designate knew the owners' total compensation. Thus, there loomed the decisive moment when the four new leaders would learn the number of zeros in the retiring owners' total compensation, revealing just how much money they had been extracting from the firm over the previous years. One or more might even question whether it was justified.

The moment did occur, and the succeeding owners ultimately decided to be more generous with the staff. None of the four was accustomed to their predecessors' high level of wealth; they didn't feel a need for it, and they wanted to distribute it in what they believed to be a more equitable manner. Part of the problem mitigated itself because the ownership portion of profit distributions now had to be divided among four instead of two, and the new leaders subsequently considered further broadening the firm's ownership base.

When Leadership Development Is Lacking

A third common obstacle to leadership succession occurs when the next generation has not been trained in business (versus project) financial management, skills that are necessary to running a firm. Although prospective firm leaders may be aces at managing projects, leading an entire firm means being able to interpret and quickly act upon a profit and loss statement, a balance sheet, or cash flow report, and to nurture relationships with lenders, meet the payroll, and manage debt. These skills must be learned; the strategic plan can provide for the development of business management acumen for prospective leaders.

STRATEGIC PLANNING AS A CATALYST FOR LEADERSHIP SUCCESSION

As key issues of a firm are identified, the strategic planning process can bring to light and put into perspective the need for leadership succession. This may precipitate the transition occurring just slightly sooner than it otherwise would have, or it may call for a more imminent or even immediate leadership change. For example, if a firm has brought in more work than it ever planned to handle—as has happened to some firms in the recent sustained strong U.S. economy—and has grown commensurately in staff size and revenue, it may have become far larger and more complex than the current leader feels at ease leading.

The kind of rapid, unexpected growth that can precipitate a leadership transition may also result from a series of strategic—or just plain lucky—hires. Talented and motivated individuals may have propelled the firm to its current level and may now be impatient with the leader and have a vision for the firm that is more ambitious than the leader is comfortable contemplating. This stress can render the leader unable to advance the firm to the satisfaction of its valuable and ambitious rainmakers. Indeed, the current leader may not want to lead, and may not enjoy managing, the firm that has burgeoned under his or her administration. But strong pride and a sense of duty to the firm may not permit him or her to express those thoughts openly. In an effective strategic planning process that includes candid discussions, this issue can be put on the table and provide the firm leader the opportunity to go out at the top of his or her game by completing a graceful, orderly, and honorable transition to the next generation.

When Strategic Planning Precipitates
Immediate Transition

A large A/E firm found itself in terrible financial shape. The firm's board of directors commissioned a strategic planning process—almost a triage. The consultants embarked on the information-gathering phase, including a detailed financial analysis and interviews with many of the firm's managers. It became increasingly clear that the firm's top leadership—both its president and chairman—bore primary responsibility for many of the firm's most serious problems, and that though no misdeeds had been committed, the firm's chief financial officer had been complicit in the firm's mismanagement. Determining that the same people who caused the problems would be unable to fix the problems, the consultants resolved to recommend the removal of *both* leaders. This wholesale leadership replacement—not transition and certainly not continuity—was a highly delicate and risky matter for the firm.

The consultants scheduled a joint meeting with the president and the chairman prior to the firm's strategic planning retreat, informing them, "We are going to recommend that both of you voluntarily step aside, in recognition that doing so would be in the best interests of the firm." Neither would be forced—or even asked—to leave the firm, but both needed to acknowledge that the firm was in need of a completely new leadership team.

The process worked. In the course of the strategic planning retreat, the firm's senior management elected the next president, and the CFO was dismissed. A well-respected former chairman of the company stepped in to serve again on an interim basis until the board of directors could elect a new, permanent chairman. Because of its decisive and courageous action, the firm was able to stanch the flow of red ink, and, over the course of the next five years, completely transformed itself. Today the firm is a large and successful leader in its target markets.

17

OWNERSHIP AND OWNERSHIP TRANSITION

Ownership in design firms is a concept fraught with misunderstandings, most often by those who do not actually have ownership—but also even by those who do. The primary source of misconceptions is the mistaken belief that ownership and leadership are connected or even synonymous. In fact, they are two entirely different subjects. *Ownership* is purely a financial investment in a firm; to be an owner, all a person has to do is own stock in the company. *Leadership* of a firm (as illustrated in Chapters 15 and 16) requires a special set of skills, business abilities, and the intangible qualities that inspire followers, which relatively few design professionals possess.

Often in the world of design firms, ownership and leadership do overlap, especially in smaller firms. At one extreme, in a sole practitioner firm, ownership and leadership are inseparable. At the other end of the size spectrum, in a much larger firm of several hundred employees or more, it is less practical to connect ownership of the firm with leadership of the firm. Think of ownership and firm leadership in terms of a Venn diagram, with two circles of varying amount of overlap depending on the firm size and composition.

Many firms provide the opportunity for select employees to purchase shares in the firm with the full knowledge that those employees have value to the firm *other* than their ability to lead the firm. An employee may be a design star, a technical guru, or an ace in construction administration, but not necessarily a firm leader, especially of the firm's business. Nonetheless, the firm wants to give those employees a vote of confidence and deepen the mutual commitment between the employee and the firm. Those valuable—but nonleader—employees who buy into their firms achieve their voice in leadership through a board of directors, elected by the shareholders, whose primary fiduciary responsibility is to act in the best interests of the shareholders.

Many design firms tend to keep the circumstances of their ownership a mystery from the staff. They rarely discuss it with nonowner employees; they often do not share the financial performance metrics of the firm except with owners; and they intentionally or inadvertently keep secret the process by which employees can become owners.

None of this mystique is necessary, and most of it is counterproductive. By disconnecting—as much as possible—the concepts of ownership and firm leadership, ownership can and should be thought of simply as the investment that it is. When a firm's management determines that an employee should have the opportunity for ownership, it is actually expressing confidence in that employee's future with the firm by inviting him or her to make a more serious and long-term commitment to the firm. And when the employee reciprocates by purchasing ownership, that employee is demonstrating his or her commitment to the firm through the hard-earned investment, but with one significant difference from other investment vehicles. When a person buys stock in a publicly held Fortune 500 company or invests in a mutual fund, he or she has no meaningful ability to influence the performance of the company or fund. The investment only has an on and off switch: the person can either leave the invested money in the stock or fund or pull it out. But when an employee invests in the design firm at which he or she works, the employee can have a great deal of impact on the success of the investment through his or her own hard work.

CHARACTERISTICS OF OWNERSHIP IN DESIGN FIRMS

Compared to many other kinds of companies in the business world, most design firms are closely held; that is, their ownership is often concentrated among a relatively small group of senior individuals who have a large—or even controlling (more than 50 percent)—interest in the company. This is very often the case for firms in which the founders are still active; in fact, it is not unheard of for a founder to maintain 100 percent ownership for most of his or her career.

Ownership in design firms typically is not very liquid. Unlike outside investments, such as stock in publicly held companies, shares in a design firm cannot be quickly cashed in through a public market or stock exchange; they must be bought back by the firm or by other shareholders, often subject to specific and sometimes restrictive terms in a shareholders' or buy/sell agreement that calls for an installment buy-back of five years or longer for large blocks of stock.

THE IMPORTANCE OF TRANSITIONING OWNERSHIP

There are five main reasons why design firms have to deal with ownership transition:

1. Founders of firms want to—and deserve to—recover the increased value of their firms derived from their own sweat equity because they "started with nothing."
2. Subsequent generations of owners want to—and also deserve to—recover the increased value of their firms derived from their initial investment when they bought in plus their own sweat equity.
3. Ownership in a design firm can serve as an incentive and powerful motivator for those who want a greater financial reward from their firm than just their salaries.
4. Opening ownership opportunities aids in retaining a firm's key employees who, without the prospect of ownership, may be tempted to move to firms that do offer that opportunity.
5. Both founders and subsequent generations of owners want to perpetuate their firms beyond their own tenure with the firm.

Considering the importance of ownership and ownership transition to the future of a design firm, a firm's strategic plan is an ideal management tool with which to address the subject.

OWNERSHIP TRANSITION ALTERNATIVES

Because of the general lack of liquidity of ownership interests in design firms, the options available to current owners—who may or may not be the founders—for transferring their ownership are somewhat limited:

- **Internal ownership transition to employees**. Most design firm owners consider internal transition—the sale of shares to some or all of the firm's employees—to be the most appealing of all options because it contributes most directly to the perpetuation and long-term stability of the firm. Also, the current or outgoing (i.e., retiring) owners can determine how large or small to make the ownership group and then handpick the incoming buyers based on whom they want to reward and motivate. Internal ownership transition comes in two forms:
  - **Direct ownership**, in which shares of the firm's stock are directly held by a few or many individual employees.
  - **An ESOP or employee stock ownership plan**, in which some or all of a firm is owned by all of the employees as a group through a qualified retirement plan using a trust vehicle. (As a rough rule of thumb, because of the initial cost of setting up and the ongoing expenses of managing such a plan, an ESOP is generally considered economically viable only for design firms larger than 50 employees.) Only in rare cases does an ESOP own 100 percent of the equity of a design firm; in most firms with ESOP ownership, the ESOP owns 50 percent or less of the firm with the remainder—most often the majority—owned directly by key employees who can drive and build the business.
- **Internal ownership transition to family members**. This option applies to very few firms, typically only those whose founder/ owner has siblings or children who want to take over and are capable of taking over the business. In such family-owned firms, the subject of ownership transition is generally "off the table" in strategic planning.
- **Acquisition of the firm**. Firm owners are often biased against being acquired because they see it as the alternative of last resort, to be pursued only if internal transition cannot be achieved. But with many strongly performing firms, the design industry

has experienced a great deal of consolidation in recent years, and selling a firm to an outside entity need not be thought of as "abandoning ship" if it achieves other objectives of the firm such as gaining increased market share, expanding geographically, obtaining "horsepower" in terms of hard-to-find personnel, providing greater opportunities for the firm's star players to grow and be challenged, or even bringing in new leadership. Furthermore, for reasons discussed below in the section on firm value, sale to an outside company almost always results in an increased return for the owners of the selling firm.

- **Close the doors.** Closing the firm *is* the option of last resort because the owner(s) receive no return at all for the value of the firm. It is understandable that in a sole proprietorship with nobody to take over the business, the single owner can only retire and refer clients elsewhere. But almost any other firm larger than one person should have created sufficient value to find a buyer, either from within the firm or from the outside.

Any of these ownership transition alternatives should be guided by the firm's long-term vision and implemented through its strategic plan.

THE VALUE OF A DESIGN FIRM

Investors in any investment vehicle—a stock, bond, mutual fund, or a design firm—expect a return from their investment. The old "buy low, sell high" adage applies in all of these situations. Likewise in a design firm, investors want the value of the firm to increase during the time that they hold their investment. Therefore, it is important to understand that the primary drivers of value in design firms boil down to financial performance. Although there may be many other important factors that the employees and owners of design firms can cite that make their firms intangibly appealing (good will, quality of projects, image in the community, work environment, awards and reputation, culture, etc.) just three main variables—its book value (assets minus liabilities), revenue size and growth, and profitability—determine what a design firm is worth to anyone inside or outside the firm looking to buy into it.

Every company has a fair market value, which is what a willing buyer would pay to a willing seller in the context of what is being offered—control in a firm transaction as in the acquisition alternative above, minority interest under a stock expansion program, or phased ownership transition. However, because the great majority of design firm owners and founders prefer to sell their stock internally to employees, they find that they have to discount the cost of their shares for two reasons: first, to make them more affordable to employee buyers whose pockets aren't as deep as those of an acquiring business; and second, because purchasing a minority stake—especially a small minority stake—in a business isn't as meaningful or valuable as acquiring a controlling interest. The level of discount applied to internal purchases usually hovers in the range of 20 to 40 percent less than the external acquisition price.

A design firm transferring shares should have a legally executed shareholders' or buy/sell agreement clearly stipulating, among other things, how the price of a share of stock or equity interest will be valued in the future. Although a firm can engage a financial/valuation advisor to determine the value of the firm each year, most firms establish a valuation formula that can be applied annually or periodically to the firm's financial performance to determine the share price for the subsequent year. These formulas, based on accrual accounting results, typically contain a number of variables that can be blended including book value, revenue, profitability, and sometimes backlog. The formula that a firm develops is important because it can be designed to incentivize the owners toward certain behavior and performance; for example, emphasizing profitability over revenue growth.

FACILITATING INTERNAL OWNERSHIP TRANSITION

Design firms have value, and buyers and sellers are entitled to have the reasonable expectation to profit from their proportional investment in that value. Yet the discount from the external sale referenced above anticipates that internal buyers—especially up-and-coming mid-level employees whose incomes haven't peaked and who may be raising families or incurring other personal expenses—are likely to have difficulty being able to afford to buy shares. Very few design firm employees have the resources to simply write a check for thousands—or tens of thousands—of dollars in order to buy an ownership share in their firm. But because current and retiring owners generally prefer to sell to insiders, many adopt methods to make the purchase more palatable:

- **Loans**. Firms that want to lower the affordability barrier for the purchase of shares can offer their employees low-interest loans for the installment purchase. The employee signs a promissory note for a given term, often three or five years, and pays back the loan directly to the firm (or the firm's bank, through a special program) over that period. If the firm lends the money at the same preferred rate at which it borrows money, the loan effectively costs the firm nothing, and both the employee and the firm gain. The buyer gains all the benefits of stock ownership including owners' profit distributions and being able to vote for the board of directors immediately upon purchasing the shares through the loan. (It would be no more realistic to expect a buyer of shares to wait for these benefits until the loan is paid back than it would for the buyer of a car to be denied use of the car until the car loan is paid off.) An easy and no-cost way for firms to make the purchase less painful is to set up the buyer's promissory note payments through automatic deductions from his or her paycheck.

- **Bonuses**. Most firms award annual or more frequent bonuses to some or all employees based on the firm's profitability. Firms can allow buyers to apply their bonuses directly toward their purchase of shares in the firm. However, firms should be careful not to increase the buyers' bonuses to such an extent that the purchase of shares feels as if it were free or a gift. Because the firm and its shares have real value, buyers are more likely to be motivated to increase that value if they have some "skin in the game" in the form of their own investment.

- **Salaries**. Quite often, when an employee is deemed worthy of becoming an owner, he or she has also achieved a threshold that justifies a salary increase. The employee can apply the incremental increase to help pay for the newly acquired shares.

STRATEGIC PLANNING TO DEVELOP A VISION FOR OWNERSHIP

Because the value of any investment, including investment in a design firm, is determined by the return on that investment a buyer can expect, a strategic plan can—and often does—establish a vision supported by goals, strategies, and action plans that aim to improve its financial performance. This creates a wonderful opportunity for synergy, because the sound business practices that every company— certainly including design firms—should adopt that make ownership in the firm financially appealing can be incorporated into its strategic plan by the very individuals in the firm who are empowered to make that happen.

During the process of developing a strategic plan, firms sometimes articulate a vision of their desired ownership philosophy, thus sending that message to all the employees and removing the veil that so often— and so unnecessarily—surrounds ownership. There is no right or wrong way to envision ownership. All that matters is what will allow the firm to be the kind of firm you want it to be.

Visions for Ownership

Here is an assortment of vision statements from real design firms that describe their philosophies toward ownership. Imagine each vision statement starting with the phrase "We will...":

- Have broad ownership as the basic building block for each owner to personally take a piece of the commitment toward excellence.
- Have reasonably broad ownership comprising a limited group of majority owners and a larger group of minority owners that provides opportunity for strategic and senior level employees—in return for performance and loyalty—to invest in the firm and take responsibility for and have a stake in its success and continuity.
- Remain broadly employee-owned, with 15 to 20 percent of employees owning shares in the company.

- Be implementing a plan to transition the firm's ownership from the founders to future leaders.
- Share our rewards though employee ownership.
- Continue broad-based employee ownership to give staff a stake in the firm's future and to share our financial success.
- Develop our ESOP culture to foster an enthusiastic, energized, collegial, professional atmosphere.
- Remain an employee-owned firm through a combination of individual and ESOP ownership.

DOS AND DON'TS FOR SUCCESSFUL OWNERSHIP TRANSITION

- **Don't start too late, especially if you are a firm founder.** Considering the funds that need to change hands, especially in a large and/or high-performing firm, you need time to develop an ownership transition plan, and buyers may need lots of time to buy your shares. The bigger your block of stock, the longer the lead time needed. Five to ten years may be a reasonable timeline to execute such a transition; otherwise, people will see that buying you out presents too steep a hill to climb.
- **Don't be greedy.** Whether you're a seller or a buyer, recognize that the future of your firm depends on your being reasonable.
- **Don't create an ownership imbalance.** Many firms that aspire to broader ownership stipulate the maximum percentage of the firm that anyone—after the founder—can own. This is not to imply that founders have done their firms wrong, but simply recognizes that embracing more employees into the ownership family helps these firms reach their vision.
- **Don't confuse ownership with leadership.** They are completely different (though they sometimes overlap in smaller firms).
- **Do engage experts.** Ownership transition is always complicated, and design professionals are trained in design, not in company valuation. Hire a small team of specialists with experience working with design firms consisting of a valuation and ownership transi-

tion consultant for strategic advice and to develop the appropriate financial mechanisms, an accountant to advise on tax ramifications, and an attorney to develop the necessary legal documents including the shareholders or buy/sell agreements and the transaction itself.

- Finally, by all means **do manage your company to be financially successful**. Otherwise, nobody—either inside or outside—will want to buy into it.

C h a p t e r

18

MARKETING

Marketing is a common term in the design industry, but few people understand what it really means. Let's define it once and for all: marketing is the first activity in the process of bringing business into a firm. The second activity is sales—pursuing and winning specific projects with specific clients. These two activities, which should be progressing at the same time in any firm, take place sequentially with respect to any particular business opportunity. Whether your firm's goal is to enter a new market or expand your share in an existing one, develop a relationship with a new client, or pursue a particular type of project, success in securing an actual project (making the sale) depends greatly on—and should be preceded by—effective marketing.

Another, less frequently used term for marketing, but one that better describes its true meaning, is *positioning*. Your marketing activities should be aimed toward positioning your firm as the go-to firm in the minds of clients, prospective clients, and anyone else who can influence client perception or serve as a source of referrals. These others include professional consultants and contractors—the people with whom you work regularly and who know your firm well—as well as the very important news media.

The term *business development* is also popular in design firms, but it is not synonymous with marketing, though the terms are often misused in that way. Business development is simply a more polished and often professionally preferred term for "sales," largely due to the aversion of some design professionals toward sales or anything related to selling.

MARKET RESEARCH

As discussed in Chapter 14, a market for design services is a group of people in the same line of business—including business in the public sector—who share a set of common priorities, concerns, and problems. Each market must be approached with a message specifically designed for it, which is much easier when that message is being directed at a group of such similar people. But before a firm can launch an effective marketing program aimed at particular markets, it must first identify target markets, determine the number of markets in which it can compete successfully, and develop a thorough understanding of each target market.

For every firm, regardless of size, there is a limited number of markets that the firm can hope to dominate and in which it can be recognized as the go-to firm by clients and prospective clients. A design firm's capacity to invest in marketing and sales activities is a zero-sum game. There are only so many marketing dollars and hours to go around, so every time a firm expends its marketing resources to position itself in a market or pursue a project, it quite literally takes away those resources from other—perhaps more strategic—positioning and pursuit activities.

Researching and validating a firm's target markets, especially when a firm plans to enter a new market, is crucial. Any investment made in marketing to penetrate a new market must be worthwhile when balanced—not necessarily equally—against investments the firm makes in other markets. Because marketing is the lifeblood of every design firm, determining the markets on which the firm will focus its energies and the relative emphasis to be placed on each is essential to a good strategic plan.

Market research consists of two components. Primary research involves collecting information firsthand from actual clients or prospective clients, either through personal interviews, telephone interviews, or client surveys. Primary research tends to result in qualitative information and provides a good sense of a market in the form of clients' perceptions, thoughts, and opinions that illuminate patterns or trends in the market. Secondary research involves collecting statistical information about the size, health, and competitiveness of a market, largely from publicly available information sources (many now found on the Internet), and results in mostly quantitative data.

CAPTURING MARKET SHARE

The overall size of any individual market for professional design services is almost always sufficiently large to support a firm that is savvy enough and committed enough to pursuing a specific market effectively, whether the market is residential developers, state departments of transportation, commercial office developers, or municipal water and sewer authorities. A firm's ability to capture a profitable market share is far more significant than the size of the market as a whole. Analyzing the number of "providers" of design services in the markets in which most design firms compete shows that they are highly fragmented, with a great many firms each obtaining a minor market share and rarely a single dominant player. The opportunity to increase market share illustrated by this simple fact is often overlooked by many firms: it matters far less how big the market "pie" is than whether a firm can win a big enough—and profitable enough—piece of it.

Part of market research should focus on whether or not it is possible to capture a big enough piece of market "pie." And though it may seem less than genteel to say it so bluntly, capturing greater market share, making yours the go-to firm, really means winning over clients who are currently employing *other* design firms. Positioning your firm within a market or competing for a client or project is no different than a Toyota dealer trying to entice longtime Ford owners to buy a Toyota. Even if the auto market as a whole trends downward, Toyota will do just fine if it is able to persuade enough erstwhile Ford owners to buy Toyotas and thereby increase its share of the overall market.

There are always many business opportunities, even in a down market. Consider the example of Southwest Airlines, which has remained profitable in difficult years during which virtually every other airline in the country has bled red ink. A design firm can achieve a similar degree of success in any market—up or down—if it positions itself as the dominant go-to supplier of needed services in its target markets. This is the power of consistent, effective marketing; it enables a firm to control its destiny regardless of what may be happening throughout the economy.

Strategically Entering a Risky Market

Condominium residences are considered to be more risky, in terms of professional liability, than other types of buildings. The risk is inherent: every 300-unit condominium project begins with a single owner, the developer, and ends with 300 owners and—unfortunately for the design firm—potential litigants. But is it possible to serve the condominium market profitably? Certainly it is. In this case, market research should be augmented by an internal analysis of a firm's ability to maintain quality, minimize errors and omissions, work closely with carefully selected developer clients to design buildings that meet or exceed the quality expectations of consumers, and execute well-written contracts. Developers do not like to be sued any more than design firms do. A design firm that offers to analyze, select, and design building systems that will help minimize future litigation by condominium owners is likely to be recognized as a firm that understands the needs of the condominium developer market.

MARKETING GOALS

Most firms embark on strategic planning out of ambition to exceed their current level of success or out of frustration with obstacles that impede their success. In either case, strategic planning is intended for the firm to realize currently unfulfilled potential. Effectiveness of the firm's marketing and sales program and activities is a frequent issue that bubbles to the surface during the planning process. The degree of importance attached to this issue ebbs and flows over time as a function of the health of the overall economy or a particular market sector, or the firm's own temporary prosperity. When the economy is healthy and work rolls in the door, firms tend not to identify marketing as a key issue. At other times, the effectiveness of a firm's marketing is a matter of life and death for the firm.

Marketing is a critical function for every firm. Firms cannot grow if they do not market effectively. And as we have explained in Chapter 13, growth is vital to a firm's survival. Even if a firm does not have an ambitious growth vision in terms of absolute size, growth is necessary because at any moment in time, a firm is either growing or shrinking—the status quo is virtually impossible to maintain.

THE POWER OF BRANDING

Marketing—positioning the firm in a target market—is not specific to any one client, project, lead, or even opportunity. It is a clear message directed at a defined group of people you want to have (or want to keep) as clients. Marketing or positioning is the art and science of getting a message out to your target audience that constantly reinforces the idea that a firm is the design leader, the go-to firm, the firm to which clients must send "requests for proposals" in its target markets.

Every firm has specific marketing objectives, but the underlying objective should always be the same: to become the household name, *the* brand name, in the minds of the target audience. The goal is for people in that market, when confronted with a design challenge, to think of a certain firm's name almost reflexively. The firm's name should become inextricably linked with the market itself. The firm should become a *brand*.

A brand is much easier to create for a product or consumer service than for a professional design service. Consider these examples of powerful product brands: Coke, Kleenex, Xerox, FedEx, Hertz. These names have become either the generic term (Kleenex) or verbs (to Xerox, to FedEx) for entire product or service categories, which puts their competition at a significant marketing disadvantage. If someone asks you "to FedEx" an overnight package, you will be far more likely to conjure up and contact FedEx—without even really thinking about it—than UPS, DHL, or any other overnight delivery company.

It's important to note that a brand and a logo are not the same thing. A logo—a graphic image—can be an important component that contributes to a brand, but it has to be backed up with substance: real capabilities, demonstrated expertise, and solid experience. For design firms in particular, a brand cannot be built around a logo (despite the importance that some design firms place on their logo's graphic elements) or around marketing statements and promises; it must communicate a firm's ability to identify with, address, and solve the unique problems of a single client-type market.

Because professional design services are provided to only a narrow subset of clients within the economy as a whole, brands in our industry are not as widely recognized as Coke or FedEx in the broad consumer economy, nor do they need to be. The only important objective is for a firm's brand to become pervasive in its target markets. A few very successful design firms have developed or become brands in their target markets. We've previously mentioned the example of HOK Sport+Venue+Event, more commonly known by its original brand, HOK Sport. The average consumer may never have heard of the company, but any member of the select group of potential clients involved in stadium design and construction certainly has. Another example is the engineering firm T.Y. Lin International, a civil and structural engineering firm specializing in the planning, design, construction engineering, and inspection of transportation infrastructure. When many state departments of transportation around the country need a long-span bridge, they think "T.Y. Lin." That greatly increases the odds of T.Y. Lin being on the RFP list for those agencies, often even before the firm takes any action to pursue a particular project. An effective brand results in the clients in a target market doing the marketing for, and perpetuating the positioning of, the branded firm.

In another realm of professional service, clients seeking expertise in sustainable design when that expertise first emerged as a specialty—before it became mainstream—would reach out or be referred to Croxton Collaborative Architects, P.C. (New York) and William McDonough and Partners (Charlottesville, Virginia). This example is instructive, because it is both typical and anomalous. "Sustainable design" does not fit our definition of a market; it is a service or expertise that can be applied to a wide range of projects for a very diverse range of clients. But these pioneering design firms recognized that there were enough clients with enough commonality of interest in this type of expertise—the very definition of a "market"—to enable them to gain market share around their specialized knowledge base and service capability. Hanging on to their dominant position, however, will likely prove difficult as more and more competitors enter the market and sustainable design becomes less and less specialized. Although they may continue to enjoy the residual advantage of having been first to market, as expertise in sustainable design becomes more widespread and even ubiquitous, these firms will have to find new ways to differentiate themselves from the competition.

THE RELATIONSHIP BETWEEN MARKETING AND SALES

If a firm has an effective marketing program, and one that preferably includes strong and unique brand identification, then it is in a better position to close sales, which is the act of getting real projects from actual clients. Selling does not come easily to many professionals in any industry, and design professionals are certainly no exception. Very few architects, engineers, or planners obtained professional degrees with the intent of becoming salespeople. Relatively few have a "sales personality," which may be a good thing for the design professions. But because of this general aversion to sales, design firms often fall back on the comfort of less personal marketing through the design of brochures and websites, neither of which involves the oft-dreaded actual contact with potential clients. However, and perhaps ironically, responsibility for sales is a natural consequence of becoming a successful architect, engineer, or planner. Advancing in a firm often means taking on more responsibility for generating business. Because selling is not a proclivity of most design professionals, effective marketing is all the more important for design firms so that whatever sales effort they can muster will have as much traction as possible. An effective marketing program makes it easier to close sales; it makes the phone ring and reduces the amount of time needed to identify and develop leads.

MEASURING THE EFFECTIVENESS OF MARKETING AND SALES ACTIVITIES

Measuring the effectiveness of marketing/positioning activities can be more of an art than a science, and most design firms rely on an intuitive approach to determining marketing success, but it is still possible to measure them in a meaningful way. A firm can calculate, for example, the number of opportunities that come in the door in the form of unsolicited requests for proposals, or less formal unsolicited requests to submit qualifications, express interest, or attend a client meeting or interview. These opportunities are the prospective projects for which a client seeks out a particular firm, and not the projects or clients that the firm pursues directly. If a firm measures the number of opportunities that come its way before launching a marketing program or a particular marketing activity, and again after the program or activity is executed, then it should be able to calculate the difference and gauge whether that program or activity has been valuable. Armed with that information, the firm can decide how to spend marketing dollars most effectively in the future.

When it comes to setting quantifiable strategic planning goals, a firm can also easily measure effectiveness at closing sales at two stages: from proposal to short list, and from short list to win. The first is easy: simply keep track of the number of proposals your firm sends out and the number of times your firm is short listed, which usually means being invited for an interview. If a firm's success rate in making the short list is lower than the firm would like it to be, then the firm may be sending out too many indiscriminate proposals, it may be producing subpar proposals, it may not be well positioned as the market's go-to firm, or there may be overt or subtle doubts about the firm's image or capabilities. The firm may need to look at the opportunities it is pursuing and its proposals with an objective, critical eye. It is also very worthwhile—if not mandatory—to obtain client feedback. Approach clients professionally and without the least hint of resentment for not being selected, and most will be happy to give an objective assessment of your proposal or presentation. Accept whatever they tell you at face value. Do not argue with or dispute it, even if you think the assessment, judgment, or selection process was unfair. Clients' perceptions—like our own—are always 100 percent valid.

Honest client feedback—sometimes in the form of a post-competition debriefing—is invaluable. There may be something about your firm's image, its proposals, or even the way its people present themselves that is more of a liability than an asset, and you may not even know about it. Moreover, simply requesting the client's thoughts and initiating that discussion helps continue rather than terminate the relationship, and may lead to an invitation to compete for that client's next project. And if, as a result of the inquiry, you conclude that the selection process was, indeed, unfair, then you have still learned something valuable: don't waste time submitting qualifications for a future project with that client!

Because clients hiring design firms are really hiring people, often for long-term engagements, nearly every project sales opportunity includes some sort of interview. By measuring the number of interviews your firm attends each year versus the number that convert to sales (projects), you can determine the effectiveness of your key employees' presentation skills. If many opportunities come your way as a result of an effective marketing program and you go on many interviews but secure few new projects, you may need to look at your interview and presentation skills closely. A design professional with a great marketing program but poor communication or social skills will not be effective. To increase your short-list-to-selection hit rate, you may need to implement a training program in public speaking for your presenters, or you may need to choose another lead presenter for your interviews. The same advice applies to interviews as to proposals with respect to client feedback. For every interview that does not lead to a job, contact the client for an honest assessment of your presentation, and use that information to improve future presentations.

ELEMENTS OF A MARKETING PROGRAM

If marketing is a key issue, some of the marketing/positioning activities to analyze as part of your strategic planning process could include the following:

- Direct mail
- Published articles in *client* (not design peer) publications
- Seminars and speeches before *client* audiences
- Public relations and communications programs

Each of these is an important element of a successful marketing and positioning program. The focus of each element is to help build recognition and drive demand for the firm, to support your objective of establishing your firm as the go-to firm in each of its target markets.

Together, they lay a strong foundation for sales by reinforcing the brand, building name recognition and reputation, and establishing the firm as the expert in a target market.

Direct Mail

To be effective, direct mail has to meet several criteria:

- It must be frequent enough to provide continual reinforcement of identity.
- It must not be overtly self-promotional.
- It must contain information that is so valuable to the recipient that he or she does not simply keep it, but *circulates it to others* within their organization—the litmus test of the value of the information.

Unfortunately, most design firm newsletters and postcards are packed with overtly self-promotional information ("Look at our project!" "Look at whom we've promoted!") that is of little interest—or more importantly, value—to clients and potential clients. Although pride in projects and people is understandable, your firm should examine its own newsletter and other direct mail pieces and analyze the percentage of space devoted to these types of topics compared to subject matter a client would find worthy of routing around their own office.

Personal Letters

A short, mass-mailed letter on your firm's letterhead that contains information of interest and value to the recipient can be a highly efficient form of direct mail. All such personal letters should be limited to one page to help ensure that they are actually read, and should address a single topic or issue. They should consist of a quick, informal discussion of a topic that makes the writer—and by extension, the firm—appear as an expert on that topic. Personal letters are inexpensive and take very little time to write and sign. They require no graphic design beyond your letterhead, and have minimal production costs for printing and mailing.

No E-mail!

Despite the increasing ubiquity of e-mail, we do not recommend sending personal letters, newsletters, or other direct mail by e-mail. Recipients' e-mail inboxes are much more cluttered than their U.S. mailboxes. E-mail is too easily caught in spam filters or immediately deleted by the recipient and does not convey the same degree of care, attention, and value as a printed and mailed piece. Even the transient marketing value of your letterhead envelope sitting on your recipients' desks—sometimes for several days or longer until they get to it—is worth the small cost of first-class postage. Every time someone is reminded of your firm's name, it has marketing value.

Published Articles

Articles written by your firm's experts on topics of interest to clients and prospective clients and published in client-oriented publications—that is, publications that your clients read—go a long way toward positioning your firm as a market leader. Published articles work best in tandem with direct mail. When a piece is published, obtain reprints and mail them directly to clients and prospective clients to draw their attention to the article. Recipients may be subscribers to the original publication, but with so many publications piled on so many office credenzas, it is unlikely that everyone on your mailing list will have focused on the article when it was originally published. The odds that a recipient will read it are much higher if the article is personally addressed to the recipient, ideally accompanied by a handwritten personal note.

Seminars and Speeches Before Clients

Sharing thoughts and ideas in front of a group of clients is an extremely potent way of establishing market leadership. However, most people, including most design professionals, fear public speaking. Even among people who have no such fear, few are effective public speakers. For design professionals, it is a skill well worth developing. It is very powerful to have the opportunity to stand before a group of prospective clients with something worthwhile to say. It conveys a strong yet subtle message of authority and leadership.

Every design firm has a few members who are effective speakers, and more firms should actively cultivate the skill among their staff. Many organizations such as Toastmasters International are specifically dedicated to the development of public speaking skills, and the cost of joining most of them is eminently affordable. And as with the development of any skill, there is no substitute for regular practice.

Professional audiences of peers, and not clients, provide an excellent environment for training. Begin with presentations within your own firm and then before small local groups, and work your way up to larger regional or national audiences.

As part of a marketing strategy, or from an overall strategic planning point of view, however, be sure not to confuse public appearances before professional audiences of design firm peers with true marketing. Sharing your knowledge with colleagues is a worthwhile and noble activity that makes the entire profession stronger, but it will not lead as directly to new business, new clients, or new markets as a presentation at, for example, the annual conventions of the Society for College and University Planning (SCUP), the Building Owners and Managers Association International (BOMA), or the Society of American Military Engineers (SAME). It is vital to take that step—which for many is a step beyond their comfort level—of speaking to audiences of potential clients. If strengthening your marketing program is among your strategic planning goals, one of the quantitative goals could be to secure a certain number of speaking engagements by a certain number of senior firm experts before a certain number of client audiences.

As an example, one architecture firm that specializes in residential design for large-scale homebuilders managed to secure speaking engagements for six of its firm members at an annual convention of the National Association of Home Builders (NAHB), one of the largest conventions in the country in any industry, with an annual attendance of more than 115,000 people. For any firm, a single speaking engagement at such an event would be a significant accomplishment. Six opportunities to present seminars before an audience of prospective clients as well as the promotional value of all convention attendees seeing the firm's name listed as recognized experts in the convention program are well worth the cost and time of preparing for and attending such an event.

Leveraging Public Relations and Communications

Relatively few design firms have sophisticated, professional public relations and communications expertise in-house. For smaller firms, it is a matter of cost; not many firms can afford an in-house public relations and communications specialist. But even for large firms, attracting and keeping the right type of expert can be difficult. PR and communications specialists are professionals just like design professionals. The best way for your firm to avail itself of the best, highest-quality expertise may be to seek and retain an outside expert. When firms try to handle PR and communications internally, they often miss the mark, because the in-house expert does not have the autonomy to act with independent professional judgment. They have far less success than they should, because the firm's leaders do not understand the full breadth of an effective PR and communications program and are unable to grasp the nuances of dealing with the media. Just as there are go-to design firms, there are go-to PR and communications firms that specialize in working with and representing design firms to the media that the clients of design firms read. In this case, *you* are the prospective client; go out and find them, evaluate them, and choose the consultant that you believe will best meet your firm's needs.

Chapter

19

EMPLOYEE RECRUITING, RETENTION, AND DEVELOPMENT

Design firms are unlike businesses that have inventories and significant capital equipment. In design firms, the principal assets are the knowledge, skills, and relationships of their employees. Like any business, design firms need to protect their assets—their people and intellectual capital—by having effective strategies for attracting, retaining, and fostering the professional development of employees.

RECRUITING AND RETENTION

Together, employee recruiting and retention address the quantifiable aspects of a firm's staffing needs. When the economy is healthy and employees are in demand, finding and keeping employees is difficult. The scarcity—and mobility—of qualified talent can become a very real constraint to business growth. But just as marketing in highly competitive markets is about positioning your firm as the *go-to firm in the minds of clients*, recruiting and retention is about attracting and retaining a significant share of the available talent—making your firm the *come-to firm for employees*. At their most basic, recruiting and retention are really marketing and sales activities. The difference is that your target audience is composed of prospective or current employees instead of prospective or current clients. And just as marketing (positioning) and sales are two sequential but interrelated activities, so should your recruiting and retention program consist of two parts: building an image of your firm as an attractive employer and then consistently and effectively closing the sale with prospective candidates and even your own employees so that they continue to want to come to work for your firm every day.

COMMUNICATING WITH CURRENT AND PROSPECTIVE EMPLOYEES

Recruiting and retention programs should constantly remind the employment pool—both internal and external—what a great place your firm is to work. That presumes, of course, that your firm actually *is* a great place to work; no recruiting and retention program can overcome an unappealing culture or work environment. If a program fails, it can be attributed to one of two possible causes: either it has not been designed or implemented effectively, or the substance is lacking.

Many of the same techniques used for marketing and positioning (see Chapter 18) can be applied to a firm's recruiting and retention. For example, a firm could maintain a database of all prior desirable job applicants and former employees, and mail them a periodic newsletter or personal letter that keeps them informed of a firm's activities, thereby letting them know of the firm's continuing interest in them.

For current employees, an effective communications strategy is also vitally important. Taking a job with a firm is no different, psychologically, than making a large consumer purchase. Employees want validation—both immediately upon hire and on a continuing basis—that they have made a good decision to join and remain with the firm. This is especially important in a competitive labor market in which recruiters are constantly and aggressively barraging employees with job offers. Many firms take it for granted that employees enjoy—or at least tolerate—what they are doing and are satisfied in their jobs. But that is never enough. Firms need an ongoing communications program to keep reminding employees how good they have it.

Thinking Like an Employee

Just as thinking from a client's point of view helps with marketing and positioning, thinking from an employee's perspective helps with recruiting and retention. The goal of most employees is to achieve the highest possible level of professional—and for most employees, financial—success. It's not selfish, egomaniacal, or greedy; most have families to think about, education loans to pay, elderly parents to care for, and dreams of owning a home and taking nice vacations. In fact, their personal goals and concerns are probably not much different from your own at a comparable stage in your career. If you wish to recruit and retain talented people, you need to give them more reasons to work for your firm than for your competition or to spin off on their own.

One way to communicate how good your employees have it is to let them appreciate the total value of their employment. At least annually, provide employees with a comprehensive statement of the total value of their compensation and benefits. Base salary is only one part of an employee's total compensation package. When the entire package is summarized on paper, it will include such things as the firm's contribution to a 401(k) or retirement plan; premiums for health, life, and disability insurance; a firm's contributions to the employee's Social Security; education and training expenses; and the value of their paid vacation, personal leave, or sick leave. The total figure can make quite an impression on an employee. This annual statement is an opportunity to let employees know the true value that the firm places on them.

Reviewing Performance Regularly

Regular and meaningful performance reviews are another important component of an employee retention program. Far too few design firms do an adequate job of it. Their untrained managers reluctantly improvise their way through performance reviews only once annually, if that often, and more often than not, the exercise is perfunctory, meaningless, uncritically complimentary, or unconstructively critical.

Employees walk away unsatisfied, unhappy, demoralized, or worse, with no idea of what they need to do to be successful in the firm. This is a tragedy, because the structure of performance reviews offers such a great opportunity for meaningful exchange and personal attention. A performance review can be awkward for both the employee and the supervisor if the supervisor has little experience conducting reviews and handles it clumsily. This is especially the case when a supervisor has an aversion to any sort of confrontation, however constructive. But this is an argument for more frequent, not less frequent, performance reviews. As with any other skill, supervisors will get better with experience. They can also benefit from formal review training, which more firms are providing.

Effective performance reviews can reinforce an employee's value, provide constructive criticism, chart a path for future career development, and make that employee even more valuable to the firm. Semiannual or even quarterly reviews are not outside the realm of possibility for any firm, regardless of size or workload. Regular reviews are particularly important for junior-level employees, who need clear information about performance expectations, prompt constructive criticism, ongoing positive reinforcement, and clear guidance for career development and advancement.

BENCHMARKING YOUR FIRM

Abundant annual statistical information about design firms is readily available from most professional associations in the industry, and from private consulting and research firms. It is useful to periodically compare your firm's employee benefits and policies against the rest of the industry to confirm that they are competitive and, at the very least, equal to industry averages. Remember that you get only what you pay for, so if your firm's goal is only to be equal to the industry in terms of benefits and policies, it is unlikely that your firm will gain "market share" by attracting more employees than your firm's competitors. Every two or three years is a good interval for this type of benchmarking, unless your firm or the employment market is changing, or you sense that something is amiss.

Looking Both Inside and Outside
the Design Industry

Be attentive to creative recruiting and retention ideas from both inside and outside the design industry. Plenty of creative ideas are widely shared that can be applied to your firm. *CE News* and *Structural Engineer* magazines, for example, conduct annual "best engineering firms to work for" competitions that include a great deal of information about what employees of engineering firms value in their employer. Firms entering the competition obtain averaged statistics from all firms entering, a nice benchmarking opportunity.

Valuable insights from other professionals and industries can be gleaned from business periodicals such as *Fortune*, which conducts its own annual competition of "The 100 Best Companies to Work For," as well as *Inc.* or *Fast Company* magazines.

Despite common misperceptions by managers, compensation is not always the top criterion in employee satisfaction. Other factors, such as the relationship with a supervisor, a flexible work schedule, job responsibilities, quality of the available work, and recognition through career advancement opportunitties may be equally important to employees, and by comparison, these things cost the firm little or nothing. A firm that understands what its employees value knows how to reward them appropriately. Conducting an employee survey every few years, as described in Chapter 3, is an excellent method of better understanding your employees' priorities and measuring your firm's progress in addressing them.

STAFF TRAINING AND DEVELOPMENT

Despite a general acknowledgement of their importance, many firms fall woefully short in staff development and training. These areas are often subjects of discussion and occupy substantial space in the back of firm leaders' minds, but they are rarely given the priority they deserve. Compared to the immediate financial pressures of maintaining high utilization—especially if that is a gauge of managers' performance and an element in an incentive compensation program—it is an unusual firm that adequately budgets the necessary cost and time for training and development, assigns principal-level attention to ensuring its success, and goes beyond the conventional and often unappreciated measures of design software technical training or "brown bag" building product sales presentations.

Most employees will seek out or obtain on the job much of the survival-level technical knowledge that they need. An effective professional development program should also focus on the "soft" skills that employees are more likely to be lacking and that are more difficult to acquire: leadership development, project management, client relationship management, communication (both written and verbal), financial understanding and management, marketing, and business development. These are the skills that future managers and leaders most need to develop. When a firm conspicuously invests in the professional development of its own staff, it sends a very positive message to both current employees and prospective job candidates that it values its employees, is committed to their personal and professional development, and considers their development to be integrally connected to the success of the firm.

20

OPERATIONAL ISSUES

Although many firms face issues that they need to address regarding their operational functions, not all of these issues require the level of analysis and collective brainpower available at the strategic planning level. When operational issues—those relating to the delivery of a design firm's services—significantly affect the firm's performance or future direction, however, they do demand the attention of the strategic planning team. Some of the more common operational issues that a firm may need to address in its strategic plan relate to project management and the project delivery process, or information technology.

PROJECT DELIVERY AND PROJECT MANAGEMENT

Two significant trends in the design and construction process related to project management that may affect design firms at the strategic level have to do with alternative project delivery methods and production outsourcing. Although they seem to come in and out of fashion and are better suited for some markets and for certain project types than others, project delivery alternatives to the traditional sequential design-bid-build model, such as design-build, demand that a firm adjust its project management and delivery approach. In these types of packaged contractual arrangements, the roles of the architect, engineer, contractor, and owner are different from what the design profession has become used to; the trend has even spawned a niche in the industry for firms that serve as project manager, acting on behalf of an owner for the entire design and construction of a project, or program manager for a series of projects.

Alternative Project Delivery Methods

The alternative project delivery methods can be thought of as contracting processes in which a product that includes a service is delivered, and the business goal of the single design-build entity, for example, is to earn a profit by minimizing costs and risk and maximizing revenue. This requires project managers to have a different set of priorities and a business and project budgeting and scheduling mind-set from the conventional approach that evolved from the concept that an architect's or engineer's duty is to protect the interests of the owner, a precedent long established throughout the design professions. In the alternative delivery scenarios, the design-builder assumes a greater risk in the project and therefore stands to benefit from a concomitant increase in potential reward, just as general contractors do in their business models.

Design-build. As design-build becomes a more common project delivery method, whether the project is designer-led or, more commonly, contractor-led, conventional roles and rules no longer apply. When compared to traditional design-bid-build project delivery, design-build involves a different team structure, a different array of players, significant differences in reporting lines and accountability, and a different set of priorities. The differences place a significant burden on design firm project managers and introduce a new type of liability for the design firm. For example, design professionals traditionally have become accustomed to protecting the interests of the owner/client of a project, particularly if those interests conflict with these of the contractor. But in design-build, the design professional may be contractually obligated to protect the interests of the contractor, who is now the firm's design-build partner or, more likely, direct client. It is important for the design firm's leadership to understand and accept the firm's role in the project, and for project managers to understand the change of roles and to execute the design firm's responsibilities accordingly.

Engineer-procure-construct and turnkey. The engineer-procure-construct (commonly known by its initials, EPC) and turnkey models are most applicable to engineering firms specializing in infrastructure or equipment-intensive facilities, but because they consolidate responsibilities in a single entity, they are likely to become more popular with clients for other kinds of projects as well. One example is a power plant in which the engineering firm designs the plant, purchases the major pieces of equipment, and installs and commissions the equipment. In a turnkey project, an electrical engineering firm or design-build team might be responsible for designing, building (including all related procurement), and bringing online an entire power plant. The client does not assume ownership or become actively involved in the management of the facility until the plant becomes operational. The engineering firm might even be responsible for hiring and training the client's facility maintenance and operations personnel.

Design-build-operate. Design-build-operate or design-build-operate-maintain projects go one or two steps further: the original design-build contractor operates the facility for the client under a long-term contract, and either retains or transfers ownership of the facility to the client at the conclusion of construction or at the end of a long-term operating contract.

The Paradox of the Project Manager

The skills and interests that lead people to become architects, engineers, and planners are not necessarily the skills and interests that make a good project manager. Effective project management involves a combination of analytical problem solving, effective people management, multitasking, and a heavy reliance on and affinity for numbers.

Technical design professionals are more likely to acquire project management skills through both formal and on-the-job training rather than through their college curriculum.

Despite the frequent lack of formal project management education typically afforded to project managers in design firms, becoming a project manager is a step on the career ladder for many design professionals, one that firm leaders use to evaluate performance, competence, and suitability for advancing to more senior management positions. The rare person who ascends to a position in senior firm leadership without having been a project manager usually has a background as a marketing principal, human resources director, or chief financial officer. The common wisdom is that if a design professional makes a project manager of the business aspects of a project, he or she could possibly also be a good manager of larger business units or an entire firm (provided that the person also possesses other qualities important to effective leadership). Therefore, aside from most design firms' need for more and better project managers to actually manage their projects, training up-and-coming technical professionals to become effective project managers is indeed strategic.

Outsourcing

Some design firms choose to outsource production tasks when they can subcontract them to firms with lower labor costs, most often overseas. There is currently more curiosity, concern, and talk than action in the design industry related to outsourcing, but as technology makes outsourcing easier, and economic pressures make it more advantageous, the point when it will become mainstream is visible on the horizon. For some firms, outsourcing is a standard practice, and for many others it may prove inevitable. The principal thesis in Thomas Friedman's *The World Is Flat: A Brief History of the Twenty-First Century* is that lower-skilled production work will inevitably migrate to places in the world where the cost of doing the work is lower, and this process is greatly facilitated by technology.

Despite concerns over liability, work quality, communication, and even the ethics of shipping technical/professional jobs out of the country, increasing numbers of U.S. firms are outsourcing construction document production to countries such as India, the Philippines, or others in Asia, where their production work can be done at night while U.S. design firm workers are at home, or to Latin America, where the production firms are in the same time zones as the United States to facilitate real-time communication. Still some other firms located in major metropolitan areas with high costs of living are outsourcing work to Canada, where exchange rates may be favorable, or even to other regions of the United States where the cost of living (and hence professional salaries) is lower. Little statistical data is available as to the scope of this activity, but anecdotal evidence suggests that there are currently more foreign firms offering these services than U.S. firms availing themselves of them, meaning that the resources are available should U.S. firms determine that outsourcing is a viable strategy for them.

Some design firms outsource early 3-D visualization work, a practice that enables them to deliver highly detailed design concepts to prospective clients in the proposal/selection stage quickly and at very low cost. Importantly, outsourcing these services typically does not raise the liability concerns of outsourcing construction document production, does not demand an extremely high level of interdisciplinary coordination, and requires less effort in overcoming cultural or language barriers, because most of the communication is more visual in nature.

An Outsourcing Success Story

One 300-person Southwest civil engineering firm has effectively had its own 40-person production outsourcing office up and running in Mexico for almost ten years. All of its technical employees in the Mexico office are graduate engineers, many are fluent English speakers, and many of the firm's U.S. employees are also fluent in Spanish. The Mexico office is technologically seamlessly connected to the firm's headquarters, the same as all the firm's offices; dialing Mexico is merely pushing an extension button on the phone, and there is complete videoconferencing with movable cameras for all offices. With all this investment, the labor cost of production work in Mexico is still just one-third the cost of the same work in the firm's U.S. offices.

INFORMATION TECHNOLOGY

Although advanced information technology and digital modeling have been a focus of considerable attention since first becoming prevalent in the design professions in the mid-1980s, issues related to information technology do not frequently rise to the level of strategic importance today. As technology becomes more sophisticated and ubiquitous, it becomes more and more like a utility, such as electrical power and telephone service, or a commodity such as pencil and paper—or computer-aided design and drafting (CADD). Few design firms can survive very far behind the information technology curve, yet few can push themselves very far ahead of it, unless information technology is a significant strategic tool that is key to realizing their vision. Only when a firm is intentionally or inadvertently at one extreme or the other is information technology likely to be an issue requiring strategic focus.

The increasing cost of maintaining a firm's information technology infrastructure is significant enough that it must be thoughtfully incorporated into annual operating and capital budgets, but watershed decisions that previously might have required the attention of a strategic planning group, such as whether to switch from paper to electronic document production or which software to choose for various activities, have fewer serious strategic implications for firms than they did years ago. Even for firms that deploy cutting-edge technology, the exact type and brand of technology itself is usually a tool, a means to an end, and not normally an element of strategic business focus.

This is not to suggest that information technology is not vitally important. Because everyone has it, and clients expect a firm to have it, information technology serves as a baseline and rarely provides a strategic, competitive advantage. Most firms recognize the need to maintain an effective, efficient, and up-to-date information technology system just as they maintain effective communication systems. But most specific decisions about software and hardware deployment can be left to the firm's leaders and IT managers. An IT issue that might arise during a strategic planning process could be that a local area network is too slow and needs to be replaced. That may become an action item in the strategic plan, but it requires little further attention of the strategic planning group. An example that illustrates an exception to this concept is a

30-person Midwest architecture firm that specializes in rollouts of retail stores and banks. This firm regularly plays in the same league as the "big boys," firms much larger—and beats them—by investing in technology that gives the firm a competitive edge.

Building Information Modeling

The current notable exception to the utility/commodity view of information technology is building information modeling (BIM), which many predict will cause significant business process change throughout the design and construction industry. The acronym BIM is used to describe a wide range of technologies, and is often mistakenly understood to be synonymous with three-dimensional virtual modeling technology, which may or may not be endowed with the type of intelligence that is an integral part of a building information model. The confusion arises from the unfortunate use of the word *model* or *modeling* in the acronym, which design professionals associate most closely with scaled, three-dimensional representations of the facilities they design. But in BIM, the correct core definition of "model" is the one used in the world of computing: a description of observed behavior that allows complex systems to be understood and their behavior predicted and that may be used as the basis for simulation.

Three-dimensional virtual models are at the core of BIM technology, because three-dimensional geometry is convenient for establishing many rules of behavior for the components of complex building systems. The purpose of a door or a window is to provide an opening in a wall surface; a simple beam must be supported in at least two points. It is possible to write software so that the virtual representation of these objects conforms to the rules that would govern the behavior of these objects in the real world. This basic level of functionality is properly called "parametric modeling," meaning that the virtual objects in the three-dimensional virtual model are defined by their parameters. It is a significant advance in technology over CADD, in which all objects were defined simply by vectors: lines having a start point, an endpoint, and an angle measured from a datum. But parametric modeling, by itself, is not BIM.

BIM is comprehensive, in that a true building information model would encompass all of the available information about a physical facility, including many attributes of building components that cannot be described by three-dimensional geometry. Doors and windows, for example, may have attributes related to fire rating, materials and finishes, reflectivity, and energy conservation and consumption. Beams have load-bearing characteristics based on their material, profile, and length. BIM is also meant to be interoperable, so that the information that a building information model contains can be easily and reliably shared among all parties involved in the design and construction process—the design team including subconsultants, general contractor, subcontractors, suppliers, and even equipment manufacturers and suppliers—and ideally, also the parties involved in the operation of a building facility throughout its useful life. This emerging technology is sometimes called digital collaboration. In fact, some of the more aggressive current proponents of BIM are on the construction side of the design and construction process; contractors are being encouraged to employ BIM by their subcontractors, product suppliers, and material fabricators who see cost savings and coordination simplification potential in their using a common building model.

When industry analysts refer to the "business process change" potential of BIM, they are usually referring to these last two attributes of the technology: comprehensiveness and interoperability. However, though most available software applications being touted as "BIM" have made significant advances in parametric modeling, the ability of design firms to generate truly comprehensive and interoperable building information models with these applications, for the time being, remains somewhat limited.

Still, it is worthwhile for design firms to monitor the development of this technology closely and assess the strategic business opportunities and threats that may emerge from it. Some firms are exploiting the benefits of parametric modeling to reduce cycle time, improve the services they offer, and gain a competitive advantage within their target markets. Others are already making routine use of design-to-fabrication technologies. Still others are positioning themselves to provide services to their clients throughout the course of the useful life of building facilities.

Integrated data models of facilities offer owners and operators the opportunity to manage those facilities far more efficiently throughout their life cycle. As the parties who first initiate or create that building information, design professionals are in a unique position to broaden the services they offer to their clients through effective information management.

Clearly, BIM will significantly alter standard operating procedures for project managers, perhaps more than for any other position in design firms. The traditional boundaries between project phases— schematic drawings, design development drawings, construction documents, and bidding/negotiation—may become blurred, and the expected deliverables of design firms are likely to change. BIM will also have a significant impact on staffing, relating to the qualifications of people needed on project teams and when those people are needed. The entry-level professional who now spends his or her time on mundane, repetitive drafting tasks—a traditional rite of professional initiation in design firms—will likely become a creature of the past. In this way, BIM may be another contributing technical factor toward outsourcing, as described above. In the future, entry-level employees will be expected to play a far more meaningful role in conceptual modeling and the development of viable, complex building solutions, and the professional distance in work tasks between senior project designers and entry-level employees will be far shorter.

BIM and other advancing technologies have a further impact on design firms. Most principals and even some project managers in these firms, especially those who came of age in an era of lesser technologies, are not conversant—much less proficient—in using these technologies. As such, there is a growing gulf between supervisors who cannot actually perform the work that they supervise and junior staff who do perform the work. This gap was nonexistent when all drawing was done on paper or Mylar; in those days, everyone in a firm was proficient using that basic hands-on "technology." But today and into the future, senior design firm managers are having to manage and rely more heavily than ever on staff members who can perform tasks that they themselves cannot perform, or sometimes even *understand*. These changes—to project delivery, not isolated in technology—are among the items that a firm may need to address in its strategic plan.

The following simple example of a strategic plan is for a fictitious firm named The Engineering Partnership. For the purposes of this model, The Engineering Partnership is a civil engineering firm, but the example illustrates the basic components of a strategic plan—mission, vision, goals, strategies, and action plans to address key issues—that apply to any design firm, including architecture, engineering, planning, environmental consulting, landscape architecture, or interior design firms.

THE ENGINEERING PARTNERSHIP'S 2007 STRATEGIC PLAN

Planning Group

Colvin	John
Kathryn	Cindy
Rich	Doug
Barbara	Susan

THE ENGINEERING PARTNERSHIP'S MISSION

Helping clients, communities, and employees succeed through environmentally responsible engineering and consulting services.

THE ENGINEERING PARTNERSHIP'S FIVE-YEAR VISION

In 2012, The Engineering Partnership will...

- Employ a staff of 180 people generating $20 million annually in net revenue at a sustained pretax, prebonus profit margin of 15%.
- Be a leading regional firm with four financially self-sufficient offices throughout our home state and a growing presence through a new office in a major metropolitan area in the Southeast or mid-Atlantic to attract talented staff and capture larger projects.
- Provide environmentally sound civil engineering and related services for our target client-type markets of municipal governments, state agencies, higher education institutions, and industrial clients.
- Be organized around our target markets, each led by a market leader with responsibility for the collective performance and profitability of his or her projects, and supported by discipline directors to maintain our technical leadership throughout the firm.
- Derive 10% of our revenue from nontraditional design services including financial feasibility consulting and grants assistance.
- Derive 20% of our revenue from alternative delivery systems including design-build.
- Visibly support the design and development of projects that contribute to a sustainable environment.
- Continue to invest substantially in advanced technology that gives us a competitive advantage.
- Foster a culture of creativity and teamwork that provides constructive feedback, rewards performance, and expects mutual accountability at all levels of the firm.
- Actively promote learning and professional advancement for all our employees.
- Have broad ownership to encourage our employees to personally make a commitment toward excellence and share in the firm's financial rewards.

THREE-YEAR FINANCIAL PERFORMANCE GOALS

Net Revenue ($ million)	FY 2006 Actual	FY 2007 Forecast	FY 2007 Goal	FY 2008 Goal	FY 2009 Goal
Northern Office	$5.0	$5.3	$5.7	$6.1	$6.3
Southern Office	$3.0	$3.2	$3.7	$4.0	$4.4
Eastern Office	$2.0	$2.3	$2.6	$2.9	$3.2
Western Office	$1.0	$1.2	$1.5	$1.7	$1.9
New Office	—	—	—	$ .8	$1.2
Total Net Revenue	$11.0	$12.0	$13.5	$15.5	$17.0
Profit Margin	9%	10%	12%	13%	15%

KEY ISSUES, STRATEGIES, AND ACTIONS FOR THE ENGINEERING PARTNERSHIP TO ACHIEVE ITS FIVE-YEAR VISION

Issue #1: Insufficient Revenue Growth

In recent years, The Engineering Partnership's revenue has plateaued and has not increased at a rate comparable to similar firms.

Primary Growth Strategies by Target Market

Municipal

- Expand water/wastewater design capabilities.
- Investigate strategic hire for new grants assistance program.

State Agencies

- Hire a transportation discipline leader.
- Increase environmental engineering staff.

Higher Education

- Build facility master planning expertise.
- Team with architects to focus on the growing science and technology facilities market.

Industrial

- Target Fortune 1000 firms within our region.
- Leverage municipal water/wastewater services for industrial clients.

Firm-Wide Growth Strategies

- Increase investment in marketing and sales to 7% of net revenue. (See marketing issue below.)
- Hire recruiting/staff development manager. (See recruiting issue below.)
- Investigate new office locations.

Growth Action Items	Who	When
Interview and engage recruiting firm to hire transportation leader.	John	Jan 15
Develop annual calendar and hold monthly meetings of municipal and industrial market leaders to share water/wastewater leads.	Kathryn	Feb 1
Develop in-house and outside vendor training curriculum for facility master planning.	Barbara	Apr 1
Research and compile list of potential grants assistance strategic hires.	Rich	Jun 1
Write risk assessment analysis for design-build projects.	Doug	Jul 15
Begin market research and make preliminary short-list recommendations for potential new office location.	Rich	Oct 1

Issue #2: Low Profitability

The Engineering Partnership's profit margin has consistently lagged behind industry medians.

Profitability Strategies

- Improve contract negotiation for defining scope, gross fee, and subconsultant fees.
- Develop and use project work plans at the beginning of each project.
- Promote better administration of contract scope and potential additional services.
- Reduce overhead expenses to a maximum of the industry median for similar firms.
- Increase utilization; record time accurately and charge all legitimate time to projects.
- Improve incentive compensation program to increase personal accountability for firm profit.

Profitability Actions	Who	When
Analyze overhead expenditures and prioritize potential savings.	Doug	Feb 1
Develop standard process for identifying, documenting, and negotiating additional services.	Doug	Apr 15
Develop project work plan template.	Doug	May 1
Research improvements to incentive compensation program and make recommendations to board of directors.	Kathryn	May 15
Instruct market leaders and discipline directors to improve the accuracy of how they account for indirect time utilization.	Doug	Jul 1
Hire consultant for negotiation training.	Doug	Sept 1

Issue #3: Organizational Structure

The Engineering Partnership's organization is geographically oriented and not focused on client types, and its organizational structure does not facilitate the firm's ability to develop specialized knowledge or expertise in our clients' unique issues.

Organizational Structure Strategies

- Develop a more client-centered organization with client/market-driven teams as the dominant reporting line.
- Increase the firm's capability to provide a wider range of services from more offices.
- Provide and emphasize dual career path potential for staff: technical and management. (See recruiting issue below.)

Organizational Structure Actions	Who	When
Complete the new market-driven organizational chart.	Colvin	Feb 1
Write job descriptions for market leaders and discipline directors.	Kathryn	Mar 15
Identify people for key positions (market leaders and discipline directors).	Kathryn	Jun 1
Assign employees to new organizational structure.	Colvin	Aug 15
Adapt accounting system to measure financial performance by market.	Doug	Sept 1
Implement new organizational structure.	Colvin	Oct 1

Issue #4: Marketing

The Engineering Partnership's marketing function does not support the firm's revenue growth goals.

Marketing Strategies

- Revise marketing materials and website to reflect client-centered organization.
- Increase visibility with prospective clients:
  - Initiate an ongoing direct mail program to continue introducing the firm.
  - Conduct more sales calls to develop personal relationships.
  - Increase speaking and presenting seminars for client groups and through professional organizations to build credibility.

Marketing Actions	Who	When
Meet with market leaders to develop an annual marketing plan for each target market.	Doug	Mar 1
Revise website to reflect market-driven organization.	Susan	Jun 1
Update printed marketing materials to reflect market-driven organization.	Susan	Jun 1
Examine structure of marketing and business development function and staff and make recommendations.	Rich	Aug 15
Develop calendar of speaking and seminar opportunities based on clients' professional association conferences.	Cindy	Sept 1

Issue #5: Recruiting

The Engineering Partnership's lack of recruiting success is inhibiting its growth.

Recruiting Strategies

- Treat recruiting strategically, recruiting proactively year-round.
- Continuously reinforce recruiting at the grassroots level with all employees.
- Publicize the firm's recruiting referral bonus program.
- Be more proactive and aggressive with entry-level hires at job fairs. Involve more junior staff.
- Develop relationships with school career development offices.
- Publicize open positions internally through the intranet, news-letters, e-mails, etc.
- Maximize the effectiveness of search firms. Explore long-term relationships.

Recruiting Actions	Who	When
Develop a firm-wide recruiting plan tailored to the needs of each market.	John	Mar 1
Identify target schools and set up calendar for on-campus interviews.	Barbara	Mar 15
Develop recruiting package.	John	Apr 1
Interview and get proposals from two search firms.	John	Apr 15
Improve website to target job candidates.	Susan	Jun 1

Communicating the Strategic Plan

- Distribute short e-mail to all staff summarizing retreat.
- Present strategic plan to employees at each office in a "road show" with generous question-and-answer time
- Write bimonthly strategic plan progress reports in in-house newsletter and post on intranet.

Implementing the Strategic Plan

- Distribute draft plan to planning group and make sure everyone understands their action items and agrees to dates. Adjust dates as necessary.
- Develop a calendar for standing monthly meetings of planning group members with action items.
- Create an overall to-do list by copying and pasting action items from all issues into a comprehensive list. Put all the action items into priority and chronological order, irrespective of the issue from which they were derived.
- Develop individual to-do lists for everyone with primary responsibility for an action item by copying and pasting action items from all issues into individual lists. Put action items in each individual list into priority and chronological order.
- Review plan progress at monthly meetings of planning group members with action items:
  - Keep the meetings efficient by using the overall action item list as the meeting agenda.
  - Develop ground rules for action items:
    - Miss the completion date once and immediately reschedule a new date.
    - Miss the rescheduled date and reassign the action item to someone else.
    - Always revisit the strategies to put each action item into context and determine whether new action items are warranted.
- Schedule a six-month checkup on plan progress by entire planning team.

INDEX